Pile Design and Construction
Rules of Thumb

Pile Design and Construction Rules of Thumb

Ruwan Rajapakse, CCM, CCE, PE

CCM – Certified Construction Manager
CCE – Certified Cost Engineer
PE – Professional Engineer

AMSTERDAM • BOSTON • HEIDELBERG • LONDON
NEW YORK • OXFORD • PARIS • SAN DIEGO
SAN FRANCISCO • SINGAPORE • SYDNEY • TOKYO

Butterworth-Heinemann is an imprint of Elsevier

Butterworth-Heinemann is an imprint of Elsevier
30 Corporate Drive, Suite 400, Burlington, MA 01803, USA
Linacre House, Jordan Hill, Oxford OX2 8DP, UK

Design Direction: Joanne Blank
Cover Design: Joe Tenerelli
Cover Images © iStockphoto

Library of Congress Cataloging-in-Publication Data
Application submitted

British Library Cataloguing-in-Publication Data
A catalogue record for this book is available from the British Library.

ISBN: 978-0-7506-8763-8

For information on all Butterworth-Heinemann publications
visit our Web site at www.books.elsevier.com

Printed and bound in the United Kingdom

Transferred to Digital Printing, 2010

Working together to grow
libraries in developing countries

www.elsevier.com | www.bookaid.org | www.sabre.org

ELSEVIER   BOOK AID
            International   Sabre Foundation

Contents

UNITS

fps Units	SI Units
Length	
1 ft = 0.3048 m	1 m = 3.28084 ft
1 in. = 2.54 cm	
Pressure	
1 ksf = 1,000 psf	1 Pascal = 1 N/m^2
1 ksf = 0.04788 MPa	1 MPa = 20,885.43 psf
1 ksf = 47,880.26 Pascal	1 MPa = 145.0377 psi
1 ksf = 47.88 kPa	
1 psi = 6,894.757 Pascal	
1 psi = 144 psf	
Area	
1 ft^2 = 0.092903 m^2	1 m^2 = 10.76387 ft^2
1 ft^2 = 144 in^2	
Volume	
1 ft^3 = 0.028317 m^3	1 m^3 = 35.314667 ft^3
Density	
1 lb/ft^3 = 157.1081 N/m^3	1 kN/m^3 = 6.3658 lbs/ft^3
1 lb/ft^3 = 0.1571081 kN/m^3	
Weight	
1 kip = 1,000 lbs	1 kg = 9.80665 N
1 lb = 0.453592 kg	1 kg = 2.2046223 lbs
1 lb = 4.448222 N	1 N = 0.224809 lbs
1 ton (short) = 2,000 lbs	1 N = 0.101972 kg
1 ton = 2 kips	1 kN = 0.224809 kips

DENSITY

water => 1 g per cubic centimeter = 1,000 g per liter = 1000 kg/m^3 = 62.42 pounds per cu. feet

Preface

Pile design is mostly about application of engineering concepts rather than use of elaborate mathematical techniques. Most pile design work can be done with simple arithmetic. I have provided necessary equations and concepts in a manner so that the reader would be able to refer to them with ease. All chapters are provided with plethora of design examples. The solutions to design examples are given in a step by step basis with many illustrations.

In geotechnical engineering, formulas and methodologies change significantly from sandy soils to clay soils. I made all attempts to separate equations based on soil type. Different pile types also would affect the design methods used. As often the case, one may find mixed soil conditions. I have provided necessary theory and design examples to tackle pile design in mixed soil conditions.

This book is mainly aimed at practicing geotechnical engineers, graduate and undergraduate students who are planning to become geotechnical engineers. The book should be a great addition to the civil professional engineer exam in USA or the chartered engineer exam in commonwealth countries. This book would also be of interest to structural engineers and architects who may run into piling work occasionally.

Ruwan Rajapakse, CCM, CCE, PE

PART 1
Introduction to Pile Selection

1

Site Investigation and Soil Conditions

The geotechnical engineer needs to develop an appropriate game plan to conduct a piling engineering project. Investigation of the site is a very important step in any geotechnical engineering project. The following major steps can be identified in a site investigation program.

- Literature survey

- Site visit

- Subsurface investigation program and sampling

- Laboratory test program

1.1 Literature Survey

The very first step in a site investigation program is to obtain published information relevant to the project.

Subsurface Information: Subsurface information can be obtained from the following sources.

- **National Geological Surveys:** Many countries have national geological surveys. In the United States, the United States Geological Survey (USGS) has subsurface information on many parts of the country. In some instances it can provide precise information in some localities. The USGS Web site, http://www.usgs.gov, is a good place to start.

- **Adjacent Property Owners:** Adjacent property owners may have conducted subsurface investigations in the past. In some cases it may not be possible to obtain information from these owners.

- **Published Literature:** Geologists have published many articles regarding the geological history of the United States. It is possible to find general information such as soil types, depth to bedrock, and depth to groundwater by conducting a literature survey on published scientific articles.

- **Aerial Photographs:** Aerial photographs are available from state agencies and private companies. *Google earth* now provides aerial maps for many parts of the world.

Aerial photographs can give information that is easily missed by borings. For example, a dark patch in the site could be organic material, or a different color stripe going through the site could be an old streambed.

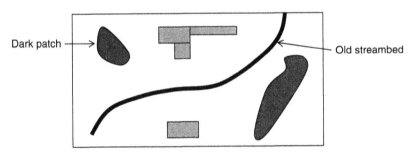

Figure 1.1   Aerial photograph

Groundwater Information: Groundwater information is extremely important during the design process.

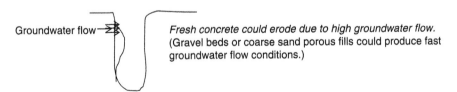

Figure 1.2   Erosion of concrete due to groundwater

Utilities: Existing utilities in the project site need to be researched and identified to avoid serious consequences. Special attention should be paid to gas and electrical utilities. Other utilities such as telephone, cable, water, sewer, and storm sewer also need to be fully and completely identified.

  The next step is to mark the utility locations in the site. A site plan should be prepared with a utility markout, indicating the type of utility, depth to the utility, and location of the utility.

If the existing utilities are not known accurately, the following procedure should be adopted.

Hand Digging Prior to Drilling: Most utilities are rarely deeper than 6 ft. Hand digging the first 6 ft prior to drilling boreholes is an effective way to avoid utilities. During excavation activities, the backhoe operator should be advised to be aware of utilities. The operator should check for fill materials because in many instances utilities are backfilled with select fill material. It is advisable to be cautious because sometimes utilities are buried with the same surrounding soil. In such cases, it is a good idea to have a second person present assigned exclusively to watch the backhoe operation.

Contaminants: The geotechnical engineer should obtain all the available information pertaining to contaminants present in the project site. Project duration and project methodology will be severely affected if contaminants are present.

1.2   Site Visit

After conducting a literature survey, it is a good idea to pay a site visit. The following information can be gathered during a site visit.

- Surface soil characteristics. Surface soil may indicate the existence of underlying fill material or loose organic soil.

- Water level in nearby streams, lakes, and other surface water bodies may provide information regarding the groundwater condition in the area.

Figure 1.3   Groundwater flow near a stream

- Closeness to adjacent buildings. (If adjacent buildings are too close, noise due to pile driving could be a problem.)

- Stability of the ground surface: This information is important in deciding the type of pile-driving rig to be used. Pile-driving rigs often get stuck in soft soils owing to lack of proper planning.

- Overhead obstructions: Special rigs may be necessary if there are overhead obstructions such as power lines.

1.3   Subsurface Investigation

Borings: A comprehensive boring program should be conducted to identify soil types existing in the site. Local codes should be consulted prior to developing the boring program.

- Typically, one boring is made for every 2,500 sq ft of the building.

- At least two-thirds of the borings should be constructed within the footprint of the building.

1.3.1   International Building Code (IBC)

The IBC recommends that borings be constructed 10 ft below the level of the foundation.

Test Pits: In some situations, test pits can be more advantageous than borings. Test pits can provide information down to 15 ft below the surface. Unlike borings, soil can be visually observed from the sides of the test pit.

Soil Sampling: Split spoon samples are obtained during boring construction. They are adequate for sieve analysis, soil identification, and

Atterberg limit tests. Split spoon samples are not enough to conduct unconfined compressive tests, consolidation tests, and triaxial tests. Shelby tube samples are obtained when clay soils are encountered. Shelby tubes have a larger diameter, and Shelby tube samples can be used to conduct consolidation tests and unconfined compressive strength tests.

1.4   Soil Types

For geotechnical engineering purposes, soils can be classified as sands, clays, and silts. The strength of sandy soils is represented with friction angle (Φ), while the strength of clay soils is represented with "cohesion."

Pure silts are frictional material and for all practical purposes behave as sands, whereas clayey silts and silty clays behave more like clays.

1.4.1   Conversion of Rocks to Soil

How did the soil originate? Geologists tell us that the young earth was made of inner magma, and at the beginning outer layer of magma was cooled and the rock crust was formed. When the earth started to cool off from its hot origin, water vapor fell onto the earth as rain. The initial rain lasted many million years until the oceans were formed. Water and dissolved chemicals eroded the rocky crust for million more years. Other factors such as meteor impacts, volcanic eruptions, and plate tectonic movements also helped to break down the original rock surface. Today, four billion years after the earth began, the first few feet of the earth are completely broken down into small pieces and are known as soils. Chemical processes are capable of breaking the larger sandy particles into much smaller clay particles.

1.4.2   Conversion of Soils to Rock

As rock breaks up and forms soil, soil also can convert back to rock. Sandy particles deposited in a river or lake with time becomes sandstone; similarly, clay depositions become shale or claystone. Sedimentary rocks originate through the sedimentation process occurring in rivers, lakes, and oceans.

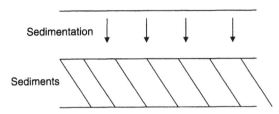

Figure 1.4   Sedimentation process in a lake

1.5   Design Parameters

1.5.1   Sandy Soils

The most important design parameter for sandy soils is the friction angle. The bearing capacity of shallow foundations, pile capacity, and skin friction of piles depend largely on the friction angle (Φ).

The strength of sandy soils comes mainly from friction between particles. The friction angle of a sandy soil can be obtained by conducting a triaxial test. There are correlations between friction angle and standard penetration test (SPT) values. Many engineers use SPT and friction angle correlations to obtain the friction angle of a soil.

To predict the settlement of a pile or a shallow foundation, one needs to use Young's elastic modulus and Poisson's ratio.

1.5.2   Clay Soils

The strength of clayey soils is developed through cohesion between clay particles. Friction is a mechanical process, whereas cohesion is an electrochemical process. Cohesion of a soil is obtained by using an unconfined compressive strength test. To conduct an unconfined compressive strength test, one needs to obtain a Shelby tube sample.

Settlement of clay soils depends on consolidation parameters. These parameters are obtained by conducting consolidation tests.

SPT — N (Standard Penetration Test Value) and Friction Angle

SPT (N) value and friction angle are important parameters in the design of piles in sandy soils. The following table provides guidelines for obtaining the friction angle using SPT values.

Table 1.1   Friction angle, SPT (N) values and relative density (*Bowles 2004*)

Soil Type	SPT (N_{70} value)	Consistency	Friction Angle (φ)	Relative Density (D_r)
Fine sand	1–2	Very loose	26–28	0–0.15
	3–6	Loose	28–30	0.15–0.35
	7–15	Medium	30–33	0.35–0.65
	16–30	Dense	33–38	0.65–0.85
	<30	Very dense	<38	<0.85
Medium sand	2–3	Very loose	27–30	0–0.15
	4–7	Loose	30–32	0.15–0.35
	8–20	Medium	32–36	0.35–0.65
	21–40	Dense	36–42	0.65–0.85
	<40	Very dense	<42	<0.85
Coarse sand	3–6	Very loose	28–30	0–0.15
	5–9	Loose	30–33	0.15–0.35
	10–25	Medium	33–40	0.35–0.65
	26–45	Dense	40–50	0.65–0.85
	<45	Very dense	<50	<0.85

Table 1.2   SPT (N) value and soil consistency

SPT (N) value vs. Total Density

Soil Type	SPT (N_{70} value)	Consistency	Total Density
Fine sand	1–2	Very loose	70–90 pcf (11–14 kN/m^3)
	3–6	Loose	90–110 pcf (14–17 kN/m^3)
	7–15	Medium	110–130 pcf (17–20 kN/m^3)
	16–30	Dense	130–140 pcf (20–22 kN/m^3)
	<30	Very dense	<140 pcf <22 kN/m^3
Medium sand	2–3	Very loose	70–90 pcf (11–14 kN/m^3)
	4–7	Loose	90–110 pcf (14–17 kN/m^3)
	8–20	Medium	110–130 pcf (17–20 kN/m^3)
	21–40	Dense	130–140 pcf (20–22 kN/m^3)
	<40	Very dense	<140 pcf <22 kN/m^3
Coarse sand	3–6	Very loose	70–90 pcf (11–14 kN/m^3)
	5–9	Loose	90–110 pcf (14–17 kN/m^3)
	10–25	Medium	110–130 pcf (17–20 kN/m^3)
	26–45	Dense	130–140 pcf (20–22 kN/m^3)
	<45	Very dense	<140 pcf <22 kN/m^3

Reference

Bowles, J., *Foundation Analysis and Design*, McGraw-Hill Book Company, New York, 1988.

1.6   Selection of Foundation Type

A number of foundation types are available for geotechnical engineers.

1.6.1   Shallow Foundations

Shallow foundations are the cheapest and most common type of foundation. They are ideal for situations when the soil immediately below the footing is strong enough to carry the building loads. In cases where soil immediately below the footing is weak or compressible, other foundation types need to be considered.

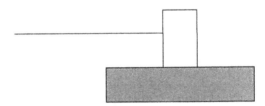

Figure 1.5   Shallow foundation

1.6.2   Mat Foundations

Mat foundations are also known as raft foundations. Mat foundations, as the name implies, spread like a mat. The building load is distributed in a large area.

Figure 1.6   Mat foundation

1.6.3   Pile Foundations

Piles are used when bearing soil is at a greater depth. In such situations, the load has to be transferred to the bearing soil stratum.

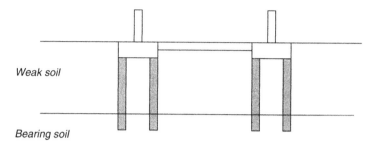

Weak soil

Bearing soil

Figure 1.7   Pile foundation

1.6.4   Caissons

Caissons are simply larger piles. Instead of a pile group, one large caisson can be utilized. In some situations, caissons can be the best alternative.

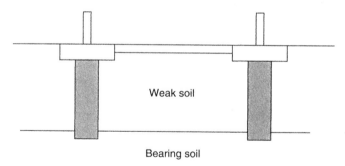

Weak soil

Bearing soil

Figure 1.8   Caissons or drilled shafts

1.6.5   Foundation Selection Criteria

Normally, every attempt is made to select shallow foundations. This is the cheapest and fastest foundation type. The designer should look into bearing capacity and settlement when considering shallow foundations.

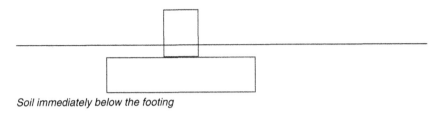

Soil immediately below the footing

Figure 1.9   Shallow foundation

The geotechnical engineer needs to compute the bearing capacity of the soil immediately below the footing. If the bearing capacity is adequate, settlement needs to be computed. Settlement can be immediate or long term. Both immediate and long-term settlements should be computed.

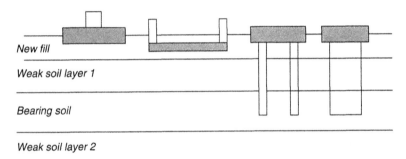

New fill

Weak soil layer 1

Bearing soil

Weak soil layer 2

Figure 1.10   Foundation types (Shallow, mat, pile and caisson)

Figure 1.10 shows a shallow foundation, mat foundation, pile group, and a caisson. The geotechnical engineer needs to investigate the feasibility of designing a shallow foundation owing to its cheapness and ease of construction. In the above situation, it is clear that a weak soil layer just below the new fill may not be enough to support the shallow foundation. Settlement in weak soil due to loading of the footing also needs to be computed.

If shallow foundations are not feasible, then other options need to be investigated. Mat foundations can be designed to carry large loads in the presence of weak soils. Unfortunately, cost is a major issue with mat foundations.

Piles can be installed as shown in the figure ending in the bearing stratum. In this situation, one needs to be careful of the second weak layer of soil below the bearing stratum. Piles could fail due to punching into a weak stratum below.

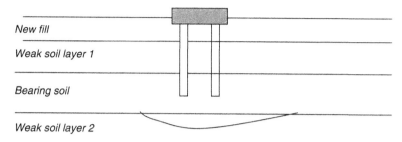

Figure 1.11   Punching failure (Soil punching into the weak soil below due to pile load)

The engineer needs to consider negative skin friction due to new fill layer. Negative skin friction would reduce the capacity of piles.

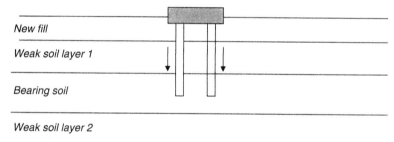

Figure 1.12   Negative skin friction due to consolidation

Due to the new load of the added fill material, weak soil layer 1 will consolidate and settle, and the settling soil will drag down piles with it. This is known as negative skin friction or down drag.

2

Pile Types

- All piles can be categorized as displacement piles and nondisplacement piles. Timber piles, closed-end steel pipe piles, and precast concrete piles displace the soil when driven into the ground. These piles are categorized as displacement piles.

- Some piles (H-piles, open-end steel tubes, hollow concrete piles).

Displacement Piles

Large displacement piles

Timber piles
Precast concrete piles
(reinforced and prestressed)
Closed-end steel pipe piles
Jacked down solid concrete piles

Small displacement piles

Tubular concrete piles
H-piles
Open-end pipe piles
Thin-shell type
Jacked down solid concrete cylinders

Nondisplacement Piles

- Steel casing withdrawn after concreting (Alpha piles, Delta piles, Frankie piles, Vibrex piles)

- Continuous flight auger drilling and concrete placement. (with or without reinforcements)

- Augering a hole and placing a thin shell and concreting

- Drilling or augering a hole and placing concrete blocks inside the hole.

2.1 Timber Piles

To have a 100-ft-long timber pile, one needs to cut down a tree of 150 ft or more.

- Timber piles are cheaper than steel or concrete piles.

- Timber piles decay due to living microbes. Microbes need two ingredients to thrive: oxygen and moisture. For timber piles to decay, both ingredients are needed. Below groundwater, there is ample moisture but very little oxygen.

- Timber piles submerged in groundwater will not decay. Oxygen is needed for the fungi (wood decaying microbes) to grow. Below groundwater level, there is no significant amount of air in the soil. For this reason very little decay occurs below groundwater level.

- Moisture is the other ingredient needed for the fungi to survive. Above groundwater level, there is ample oxygen.

- States such as Nevada, Arizona, and Texas have very little moisture above the groundwater level. Since microbes cannot survive without water, timber piles could last a long period of time.

- This is not the case for states in northern parts of the United States. A significant amount of moisture will be present above groundwater level due to snow and rain. Hence, both oxygen and moisture are available for the fungi to thrive. Timber piles will decay under such conditions. Creosoting and other techniques should be used to protect timber from decay above groundwater level.

- When the church of St. Mark in Venice was demolished owing to structural defects in 1902, the wood piles driven in AD 900 were in good condition. These old piles were reused to construct a new tower in place of the old church. Venice is a city with very high groundwater level, and piles were under water for centuries.

- Engineers should consider possible future construction activities that could trigger lowering of the groundwater level.

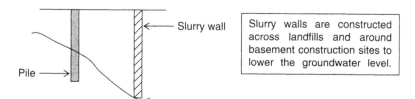

Figure 2.1   Lowering of groundwater due to slurry wall

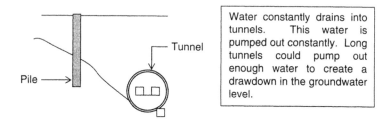

Figure 2.2   Lowering of groundwater due to tunnel

2.1.1   Timber Pile Decay—Biological Agents

Many forms of biological agents attack timber piles. Timber is an organic substance, and nature will not permit any organic substance to go to waste.

Fungi: Fungi belong to the plant kingdom. The main distinction between plants and animals is the plants' ability to generate food on their own. On the other hand, all food types of animals come from plants. On that account fungi differ from other plants. Fungi are not capable of generating their own food because they lack chlorophyll, the agent that allows plants to generate organic matter using sunlight and inorganic nutrients. For this reason, fungi have to rely on organic matter for its supply of food.

Fungi need organic nutrients (the pile itself), water, and atmospheric air to survive.

Identification of fungi attack: Wood piles subjected to fungi attack can be identified by their swollen, rotted, and patched surface. Unfortunately, many piles lie underground and are not visible. For this reason, piles that could be subjected to environmental conditions favorable to fungi growth should be treated.

Marine Borers: Marine borers belong to two families: mollusk and crustacean. Both groups live in seawater and in brackish water. Waterfront structures usually need to be protected from them.

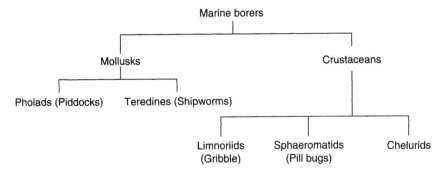

Figure 2.3   Marine borers

Pholads (Piddocks): These are sturdy creatures that can penetrate the toughest of woods. Pholads can hide inside their shells for prolonged periods of time.

Teredines (Shipworms): Teredines are commonly known as shipworms because of their wormed like appearance. They typically enter the wood at larvae stage and grow, inside the wood.

Limnoriids: Limnoria, a member of the crustacean, group has seven legs and could grow to 6 mm. They are capable of digging deep into the pile and damage the pile within, without any outward sign.

Sphaeromatids: Sphaeromatids create large-size holes, approximately ½ in. in size, and can devastate wood piles and other marine structures.

Chelurids: Chelurids are known to drive out Limnoriids from their burrows and to occupy them.

2.1.2   Preservation of Timber Piles

Three main types of wood preservatives are available.

1. Creosote

2. Oil-borne preservatives

3. Water-borne preservatives

These preservatives are usually applied in accordance with the specifications of the American Wood Preserver's Association (AWPA).

Preservatives are usually applied under pressure, hence the term *pressure treated.*

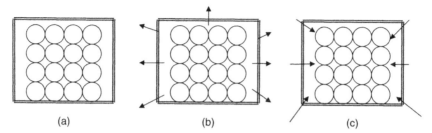

(a)   (b)   (c)

Figure 2.4   Pressure treating timber plies. (a) Timber piles are arranged inside a sealed chamber. (b) Timber piles are subjected to a vacuum. Due to the vacuum, moisture inside piles is removed. (c) After applying the vacuum, the chamber is filled with preservatives. The preservative is subjected to high pressure until a prespecified volume of liquid is absorbed by the wood.

Shotcrete Encasement of Timber Piles

- Shotcrete is a mixture of cement, gravel, and water. High-strength shotcrete is reinforced with fibers.

- Shotcrete is sprayed onto the top portion of the pile where it can possibly be above groundwater level.

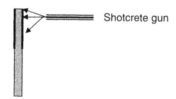

Figure 2.5   Shotcrete encasement of timber piles

2.1.3   Timber Pile Installation

Timber piles need to be installed with special care. They are susceptible to brooming and damage. Any sudden decrease in driving resistance should be investigated.

Splicing of Timber Piles

- Splicing of timber piles should be avoided if possible. Unlike steel or concrete piles, timber piles cannot be spliced effectively.

- The usual practice is to provide a pipe section (known as a sleeve) and bolt it to two piles.

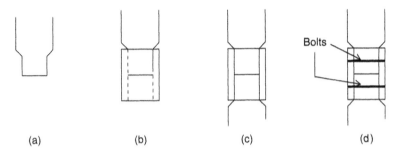

(a)   (b)   (c)   (d)

Figure 2.6   Splicing of timber piles. (a) Usually, timber piles are tapered prior to splicing as shown here. (b) The sleeve (or the pipe section) is inserted. (c) The bottom pile is inserted. (d) The pipe section is bolted to two piles.

- Sleeve joints are approximately 3 to 4 ft in length. As one can easily see, the bending strength of the joint is much lower than the pile. Splice strength can be increased by increasing the length of the sleeve.

- Most building codes require that no splicing be conducted on the upper 10 ft of the pile since the pile is subjected to high bending stresses at upper levels.

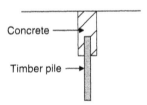

Figure 2.7   Concrete–timber composite pile

- **Sleeves Larger Than the Pile Diameter:** Sleeves larger than the pile may get torn and damaged during driving and should be avoided. If this type of sleeve is to be used, the engineer should be certain that the pile sleeve is not driven through hard strata.

- **Uplift Piles:** Timber splices are extremely vulnerable to uplift (tensile) forces and should be avoided. Other than sleeves, steel bars and straps are also used for splicing.

2.2   Steel H-Piles

- Timber piles cannot be driven through hard ground.

- Steel H-piles are essentially end-bearing piles. Due to limited perimeter area, H-piles cannot generate much frictional resistance.

- Corrosion is a major problem for steel H-piles. The corrosion is controlled by adding copper into steel.

- H-piles are easily spliced. They are ideal for highly variable soil conditions.

- H-piles can bend under very hard ground conditions. This is known as *dog legging*, and the pile installation supervisor needs to make sure that the piles are not out of plumb.

- H-piles can get plugged during the driving process.

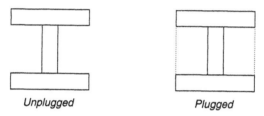

Unplugged   Plugged

Figure 2.8   Soil plugging

- If the H-pile is plugged, end bearing may increase due to larger area. On the other hand, skin friction may become smaller due to smaller wall area.

- When H-piles are driven, both analyses should be done (unplugged and plugged) and the lower value should be used for design.

  Unplugged: Low end bearing, high skin friction;

  Plugged: Low skin friction, high end bearing.

2.2.1   Splicing of H-Piles

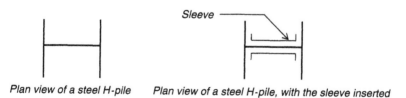

Plan view of a steel H-pile   Plan view of a steel H-pile, with the sleeve inserted

Sleeve

Figure 2.9   Splicing of H-piles

- Step 1: The sleeve is inserted into the bottom part of the H-pile as shown and bolted to the web.

- Step 2: The top part of the pile is inserted into the sleeve and bolted.

Guidelines for Splicing (International Building Code)

IBC states that splices shall develop not less than 50% of the pile bending capacity. If the splice is occurring in the upper 10 ft of the pile, eccentricity of 3 in. should be assumed for the column load. The splice should be capable of withstanding the bending moment and shear forces due to a 3-in. eccentricity.

2.3   Pipe Piles

- Pipe piles are available in many sizes; 12-in.-diameter pipe piles have a range of thicknesses.

- Pipe piles can be driven either open end or closed end. When driven open end, the pipe is cleaned with a jet of water.

2.3.1   Closed-End Pipe Piles

- Closed-end pipe piles are constructed by covering the bottom of the pile with a steel plate.

- In most cases, pipe piles are filled with concrete. In some instances, pipe piles are not filled with concrete to reduce the cost. If pipe piles are not filled with concrete, then corrosion protection layer should be applied.

- If a concrete-filled pipe pile is corroded, most of the load-carrying capacity of the pile remains intact due to concrete. On the other hand, empty pipe pile loses a significant amount of its load-carrying capacity.

- Pipe piles are a good candidate for batter piles.

- The structural capacity of pipe piles is calculated based on concrete strength and steel strength. The thickness of the steel should be reduced to account for corrosion (and is typically reduced by 1/16 in. to account for corrosion).

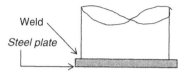

A pipe pile is covered with an end cap. The end cap is welded as shown.

Figure 2.10   Closed-end pipe pile

- In the case of closed-end driving, soil heave can occur. Open-end piles can also generate soil heave, due to plugging of the open end of the pile with soil.

- Pipe piles are cheaper than steel H-piles or concrete piles.

2.3.2 Open-End Pipe Piles

- Open-end pipe piles are driven, and the soil inside the pile is removed by a water jet.

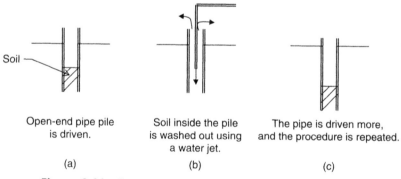

Open-end pipe pile is driven.	Soil inside the pile is washed out using a water jet.	The pipe is driven more, and the procedure is repeated.
(a)	(b)	(c)

Figure 2.11   Driving procedure for open-end pipe piles

- Open-end pipe piles are easier to drive through hard soils than closed-end pipe piles.

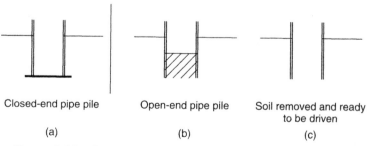

Closed-end pipe pile	Open-end pipe pile	Soil removed and ready to be driven
(a)	(b)	(c)

Figure 2.12   Open-end pipe pile and closed-end pipe piles

Ideal Situations for Open-End Pipe Piles

- Soft layer of soil followed by a dense layer of soil

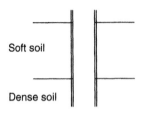

Soft soil	• **Site condition:** Soft layer of soil followed by a hard layer of soil.
	• Open-end pipe piles are ideal for this situation. After driving to the desired depth, soil inside the pipe is removed and concreted.
Dense soil	• Closed-end pipe piles also could be considered to be ideal for this type of situation.

Figure 2.13   Open-end pipe pile in soft soil underlain by dense soil

- Medium-dense layer of soil followed by a dense layer of soil

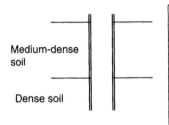

Medium-dense soil	• **Site condition:** Medium-dense layer of soil followed by a hard layer of soil.
	• Open-end pipe piles are ideal for this situation as well.
	• Closed-end pipe piles may *not* be a good choice, since driving closed-end pipe piles through medium-dense soil layer may be problematic.
Dense soil	• It is easier to drive open-end pipe piles through a dense soil layer than closed-end pipe piles.

Figure 2.14   Open-end pipe pile in medium dense soil underlain by dense soil

Telescoping

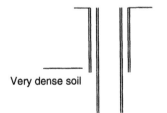

Very dense soil	• Usually telescoping is conducted, when very dense soil is encountered.
	• In such situations, it may not be possible to drive a larger pipe pile.
	• Hence, a small-diameter pipe pile is driven inside the original pipe pile.

Figure 2.15   Telescoping

- Due to the smaller diameter of the telescoping pipe pile, the end-bearing capacity of the pile is reduced. To accommodate the loss, the length of the telescoping pile should be increased.

Splicing of Pipe Piles

Pipe piles are spliced by fitting a sleeve. The sleeve fits into the bottom section of the pile as well as the top section.

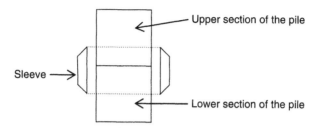

Figure 2.16   Slicing of pipe piles

2.4   Precast Concrete Piles

Precast concrete piles can be either reinforced concrete piles or prestressed concrete piles.

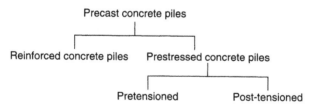

Figure 2.17   Precast concrete piles

2.4.1   Reinforced Concrete Piles

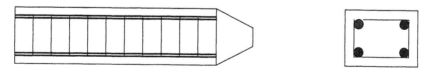

Figure 2.18   Reinforced concrete piles

2.4.2   Prestressed Concrete Piles

Pre-tensioning Procedure

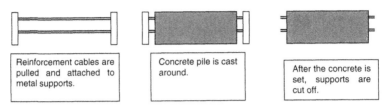

| Reinforcement cables are pulled and attached to metal supports. | Concrete pile is cast around. | After the concrete is set, supports are cut off. |

Figure 2.19   Pre-tensioning procedure

Post-tensioning Procedure

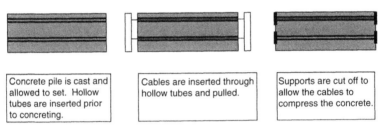

| Concrete pile is cast and allowed to set. Hollow tubes are inserted prior to concreting. | Cables are inserted through hollow tubes and pulled. | Supports are cut off to allow the cables to compress the concrete. |

Figure 2.20   Post-tensioning procedure

International Building Code—IBC: IBC specifies a minimum lateral dimension (diameter or width) of 8 in. for precast concrete piles.

Reinforcements for Precast Concrete Piles (IBC)
As per IBC, longitudinal reinforcements should be arranged symmetrically. Lateral ties should be placed every 4 to 6 in.

Concrete Strength (IBC)
As per IBC, 28-day concrete strength (f'_c) should be no less than 3,000 psi.

2.4.3   Hollow Tubular Section Concrete Piles

- Most hollow tubular piles are post-tensioned to withstand tensile stresses.

- Hollow tubular concrete piles can be driven either closed end or open end. A cap is fitted at the end for closed end driving.

• These piles are not suitable for dense soils.

• Splicing is expensive.

• Cutting off these piles is a difficult and expensive process. It is very important to know the depth to the bearing stratum with reasonable accuracy.

2.4.4   Driven Cast-in-Place Concrete Piles

• STEP 1: The steel tube is driven first.

• STEP 2: Soil inside the tube is removed by water jetting.

• STEP 3: A reinforcement cage is set inside the casing.

• STEP 4: The empty space inside the tube is concreted while removing the casing.

2.4.5   Splicing of Concrete Piles

Splicing of prestressed concrete piles is not an easy task.

• Holes are drilled in the bottom section of the pile.

• Dowels are inserted into the holes and filled with epoxy.

• Holes are driven in the upper section of the pile as well.

• The top part is inserted into the protruding dowels.

• Epoxy grout is injected into the joint.

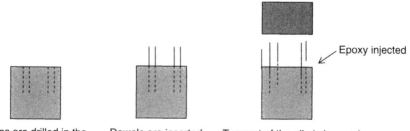

Holes are drilled in the   Dowels are inserted   Top part of the pile is lowered.
lower part of the pile.   and epoxy is injected.

Figure 2.21   Splicing of concrete piles

2.5 Casing Removal Type

2.5.1 Frankie Piles

The following major components are needed for Frankie piles: pile hammer, casing, mandrel, bottom cap, reinforcement cage, and concreting method.

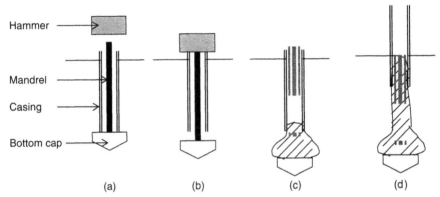

Figure 2.22 Frankie pile installation. (a) The casing is driven to the desired depth using an internal mandrel. (b) The bottom cap is broken off by heavy hammer drops. (c) The mandrel is removed, and the reinforcement cage is inserted. (d) The casing is lifted and concreted.

2.5.2 Delta Piles

The following major components are needed for Delta piles: *pile hammer, casing, mandrel, bottom cap, concreting method.*

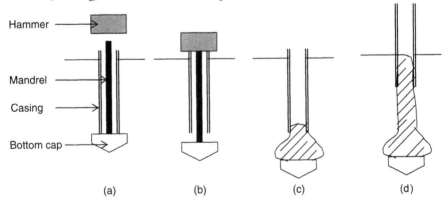

Figure 2.23 Delta pile installation. (a) The casing is driven to the desired depth using an internal mandrel. (b) The bottom cap is broken off by heavy hammer drops. (c) The mandrel is removed. (d) The casing is lifted and concreted (no reinforcement cage).

2.5.3   Vibrex Piles (Casing Removal Type)

The following major components are needed for Vibrex piles: vibrating hammer, casing, bottom cap, reinforcement cage, and concreting method.

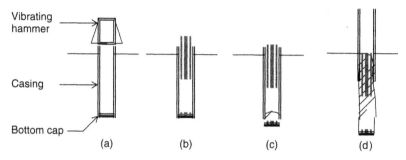

Figure 2.24   Vibrex pile installation. (a) The casing is driven to the desired depth using a vibrating hammer. (b) The reinforcement cage is inserted. (c) The bottom cap is removed. A removal mechanism is needed to remove the bottom cap. (d) The casing is lifted and concreted.

2.5.4   Compressed Base Type

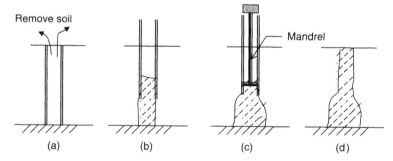

Figure 2.25   Compressed base piles. (a) Insert a casing to the hard stratum and remove soil inside the casing. (b) Lift the casing and concrete. (c) Apply pressure using a hammer and an internal mandrel. (d) Concrete rest of the pile.

In this pile type, a larger base area is created. In most cases, it is not easy to estimate the bearing area of the base.

2.6   Precast Piles with Grouted Base

Grouted base piles are constructed by augering a hole and grouting the bottom under pressure. A precast pile (steel, concrete, or timber) is inserted into the grouted base.

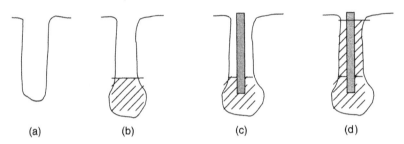

(a)   (b)   (c)   (d)

Figure 2.26   Precast piles with grouted base. (a) Auger a hole. (b) Grout the base under pressure. (c) Insert the precast pile. (d) Grout the annulus of the hole.

2.6.1   Capacity of Grouted Base Piles

Failure mechanisms of grouted base piles

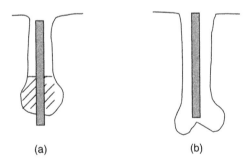

(a)   (b)

Figure 2.27   Failure of grouted base piles. (a) Pile penetrates the grouted base and fails (punching failure). (b) Grouted base fails (splitting failure).

2.7   Mandrel-Driven Piles

Theory: Mandrels are used to drive a thin shell into the soil. The mandrel is later withdrawn, and the shell is concreted.

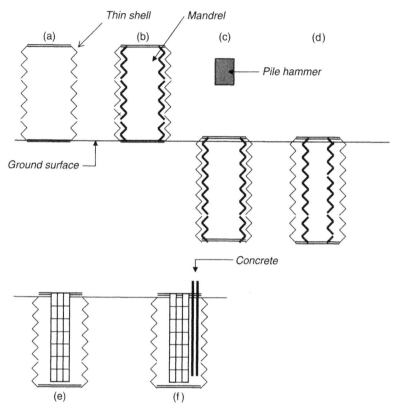

Figure 2.28   Mandrel driven piles. (a) Thin shell. (b) Insert the mandrel into the thin shell. The mandrel is fitted into the corrugations of the thin shell. The mandrel is a solid object that can be driven into the ground with a pile hammer. (c) Drive the mandrel and the thin shell into the ground. The mandrel will drag down the thin shell with it. (d) Collapse the mandrel. (Mandrels are collapsible). After the mandrel is collapsed, it is removed from the hole. (e) Insert the reinforcement cage. (f) Concrete the thin shell.

Note: Mandrel-driven piles are not suitable for unpredictable soil conditions. It is not easy to increase the length of the mandrel if the pile had to be driven to a longer depth.

2.8 Composite Piles

Piles that consist of two or more different types of piles are known as composite piles.

2.8.1 Pipe Pile/Timber Pile Composite

Method 1

It is a well-known fact that timber piles decay above groundwater. For this reason, steel (or concrete) pipe piles are used above groundwater level.

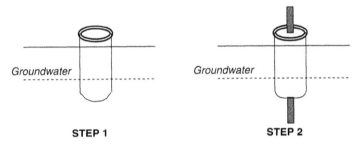

Figure 2.29 Pipe pile–timber pile composite. STEP 1: Drive the pipe pile below the groundwater level. STEP 2: Drive the timber pile inside the steel or concrete pipe pile. STEP 3: Concrete the annulus between the pipe pile and the timber pile.

Method 2

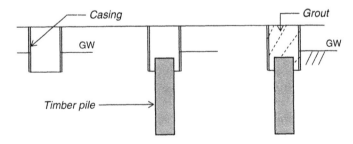

Figure 2.30 Pipe pile–timber pile composite (short timber pile).

- A steel pipe is driven into the ground, and soil inside the pipe is removed. The timber pile is driven inside the casing.

- The casing is grouted.

2.8.2   Precast Concrete Piles with H-Section

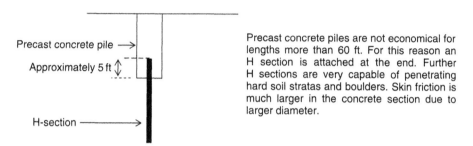

Precast concrete pile →

Approximately 5 ft ↕

H-section ⟶

Precast concrete piles are not economical for lengths more than 60 ft. For this reason an H section is attached at the end. Further H sections are very capable of penetrating hard soil stratas and boulders. Skin friction is much larger in the concrete section due to larger diameter.

Figure 2.31   Concrete pile–H-pile composite.

2.8.3   Uncased Concrete and Timber Piles

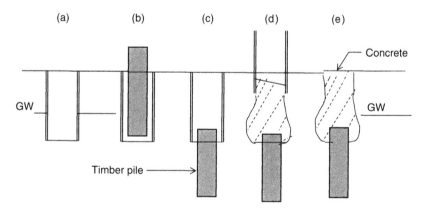

(a)   (b)   (c)   (d)   (e)

Concrete

GW

GW

Timber pile ⟶

Figure 2.32   Concrete–timber composite pile

- A steel casing is driven. The soil is removed and the timber pile is installed (Figs. 2.32a, b, c).

- Concreting is done while the casing is lifted (Figs. 2.32d, e).

- The timber pile is installed below the groundwater level.

2.9   Fiber-Reinforced Plastic Piles (FRP Piles)

- Fiber-reinforced plastic (FRP) piles are becoming increasingly popular because of their special properties.
- Timber piles deteriorate over time and are also vulnerable to marine borer attack. Plastic piles are not vulnerable to marine borer attack.
- Corrosion is a major problem for steel piles.
- Concrete piles may deteriorate in marine environments.
- For a instance, plastic pile with steel core can be used in highly corrosive environments. A plastic outer layer can protects the inner core from marine borer attack and corrosion.

Materials Used

Fiberglass and high density polyethylene (HDPE) plastic are the most popular materials used.

Types of FRP Piles:

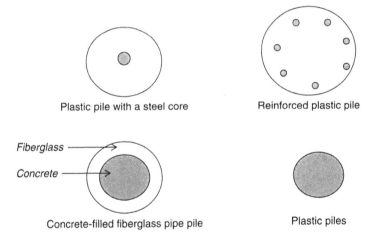

Figure 2.33   Fiberglass composite piles

Plastic Pile with a Steel Core: A steel core provides rigidity and compressive strength to the FRP pile. In some instances, the plastic could peel off from the steel core.

Reinforced Plastic Piles: Most of these piles are made of recycled plastic. Instead of the steel core, steel rebars are also used to provide rigidity and strength.

Fiberglass Pipe Piles: These piles consists of a fiberglass pipe and a concrete core. Typically, fiberglass piles are driven and filled with concrete.

Plastic Lumber: Plastic lumber consists of recycled HDPE and fiberglass.

Use of Wave Equation for Plastic Piles: Wave equation can be used for driveability analysis of plastic piling by using adjusted piling parameters.

Reference

Iskander M. G., et al. "Driveability of FRP Composite Piling," *ASCE J. of Geotechnical and Geoenvironmental Eng.*, Feb. 2001.

3

Selection of Pile Type

Once the geotechnical engineer has decided to use piles, the next question is, which piles? Many types of piles are available as discussed in Chapter 20. Timber piles are cheap but difficult to install in hard soil. Steel piles may not be good in marine environments owing to corrosion.

Some possible scenarios are discussed in the following.

CASE 1
Granular soil with boulders underlain by medium-stiff clay.

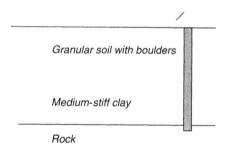

Figure 3.1   End bearing pile

Timber piles are not suitable for the situation given above owing to the existence of boulders in upper layers. The obvious choice is to drive piles all the way to the rock. The piles can be designed as end bearing piles. If the decision is taken to drive piles to the rock for the above configuration, H-piles are ideal. Unlike timber piles or pipe piles, H-piles can go through boulders.

On the other hand, the rock may be too far for the piles to be driven. If the rock is too deep for piles, a number of alternatives can be envisioned.

1. Drive large-diameter pipe piles and place them in medium-stiff clay.
2. Construct a caisson and place it in the medium-stiff clay.

In the first case, driving large-diameter pipe piles through boulders could be problematic. It is possible to excavate and remove boulders if the boulders are mostly at shallower depths.

In the case of H-piles, one has to be extra careful not to damage the piles.

Caissons placed in the medium-stiff clay is a good alternative. Settlement of caissons needs to be computed.

CASE 2

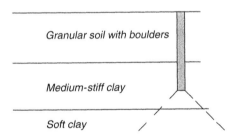

Figure 3.2   Pile ends in medium stiff clay

Piles cannot be placed in soft clay, but they can be placed in medium-stiff clay. In this situation, piles need to be designed as friction piles. Pile capacity comes mainly from end bearing and skin friction. End bearing piles, as the name indicates, obtain their capacity mainly from the end bearing, while friction piles obtain their capacity from skin friction.

If the piles were placed in the medium-stiff clay, stresses would reach the soft clay layer below. The engineer needs to make sure that settlement due to compression of soft clay is within acceptable limits.

CASE 3

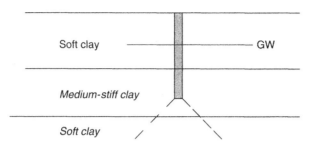

Figure 3.3   Pile ends in medium stiff clay with groundwater present

Timber Piles: Timber piles can be placed in the medium-stiff clay. In this situation, soft clay in upper layers may not pose a problem for driving. Timber piles above groundwater should be protected.

Pipe Piles: Pipe piles can also be used in this situation. Pipe piles cost more than timber piles. One of the main advantages of pipe piles over timber piles is that they can be driven hard. At the same time, large-diameter pipe piles may be readily available, and higher loads can be accommodated by few piles.

H-Piles: H-piles are ideal candidates for end bearing piles. The above situation calls for friction piles. Hence, H-piles may not be suitable for the above situation.

PART 2
Design of Pile Foundations

4

Pile Design in Sandy Soils

A modified version of the Terzaghi bearing capacity equation is widely used for pile design.

The third term or the density term in the Terzaghi bearing capacity equation is negligible in piles and hence usually ignored. The lateral earth pressure coefficient (K) is introduced to compute the skin friction of piles.

$$P_{ultimate} = \underbrace{[\sigma'_t \times N_q \times A]}_{\text{End Bearing Term}} + \underbrace{[K \cdot \sigma'_v \times \tan \delta \times A_p]}_{\text{Skin Friction Term}}$$

$P_{ultimate}$ = ultimate pile capacity
σ'_t = effective stress at the tip of the pile
N_q = bearing factor coefficient
A = cross-sectional area of the pile at the tip
K = lateral earth pressure coefficient. (Use the table below for K values)
σ'_v = effective stress at the perimeter of the pile. (σ'_v varies with the depth. Usually, σ'_v value at the midpoint of the pile is obtained).
Tan δ = friction angle between pile and soil. See the table below for δ values.
A_p = perimeter area of the pile

For round piles, $A_p = (\pi \times d) \times L$ (d = diameter, L = length of the pile)

Description of Terms

Effective Stress (σ') = when a pile is driven, effective stress of the existing soil around and below the pile will change. Figure 4.1 shows effective stress prior to driving a pile.

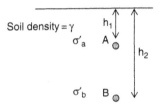

Figure 4.1   Effective stress

$$\sigma'_a = \gamma\, h_1$$
$$\sigma'_b = \gamma\, h_2$$

Effective stress after driving the pile is shown in Figure 4.2. When a pile is driven, soil around the pile and below the pile becomes compacted.

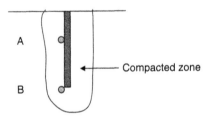

Figure 4.2   Compaction of soil around driven piles

Because of soil disturbance during the pile-driving process, σ'_{aa} and σ'_{bb} cannot be accurately computed. Usually the increase in effective stress due to pile driving is ignored.

Nq (Bearing Capacity Factor): Many researchers have provided techniques to compute bearing capacity factors. End bearing capacity is a function of friction angle, dilatancy of soil, and relative density. All these parameters are lumped into N_q. Different methods of obtaining the N_q value will be discussed here.

K (Lateral Earth Pressure Coefficient): Prior to discussing the lateral earth pressure coefficient related to piles, it is necessary to investigate lateral earth pressure coefficients in general.

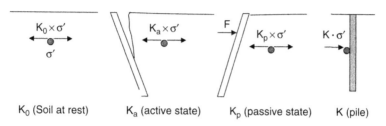

Figure 4.3   Lateral earth pressure coefficients

K_0—In-situ soil condition: For at rest conditions, horizontal effective stress is given by $K_0 \times \sigma'$ ($K_0 =$ lateral earth pressure coefficient at rest and σ'-vertical effective stress).

K_a—Active condition: In this case, soil is exerting the minimum horizontal effective stress, since soil particles have room to move ($K_a =$ active earth pressure coefficient). K_a is always smaller than K_0.

K_p—Passive condition: In this case, soil is exerting the maximum horizontal effective stress, since soil particles have been compressed ($K_p =$ passive earth pressure coefficient). K_p is always greater than K_0 and K_a.

K—Soil near piles: Soil near a driven pile will be compressed. In this case, soil is definitely exerting more horizontal pressure than the in-situ horizontal effective stress (K_0). Since K_p is the maximum horizontal stress that can be achieved, K should be between K_0 and K_p.

$$K_a < K_0 < K \text{ (pile condition)} < K_p$$

Hence, $K = (K_0 + K_a + K_p)/3$ can be used as an approximation. Equations for K_0, K_a, and K_p are:

$$K_0 = 1 - \sin\phi$$
$$K_a = \tan^2(45 - \phi/2)$$
$$K_p = \tan^2(45 + \phi/2)$$

Tan δ (wall friction angle)—The friction angle between pile material and soil δ decides the skin friction. Friction angle (δ) varies with the pile material and soil type. Various agencies have conducted laboratory tests and have published δ values for different pile materials and soils.

A_p (Perimeter Surface Area of the Pile)

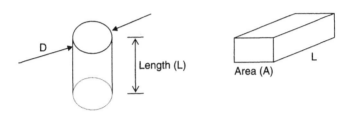

Figure 4.4   Perimeter surface area

Perimeter surface area of a circular pile $= \pi \cdot D \cdot L$
Perimeter surface area of a rectangular pile $= A \cdot L$
Skin friction acts on the perimeter surface area of the pile.

4.1   Equations for End Bearing Capacity in Sandy Soils

A number of methods are available for computing the end bearing capacity of piles in sandy soils.

API Method (American Petroleum Institute, 1984)

$q = N_q \cdot \sigma'_t$
q = end-bearing capacity of the pile (units same as σ'_t)
σ'_t = effective stress at pile tip
(Maximum effective stress allowed for the computation is 240 kPa or 5.0 ksf.)
N_q = 8 to 12 for loose sand
N_q = 12 to 40 for medium-dense sand
N_q = 40 for dense sand

Note: Sand consistency (loose, medium, and dense) can be obtained from Tables 1.1 and 1.2.

Martin et al. (1987)

SI units: $q = C \cdot N$ (MN/sq m)
fps units: $q = 20.88 \times C \cdot N$ (ksf)
q = end bearing capacity of the pile

N = SPT value at pile tip (blows per foot)
C = 0.45 (for pure sand)
C = 0.35 (for silty sand)

Example: SPT (N) value at the pile tip is 10 blows per foot. Find the ultimate end bearing capacity of the pile assuming that the pile tip is in pure sand and the diameter of the pile is 1 ft.

fps units: $q = 20.88 \times C \cdot N$ (ksf)
$q = 20.88 \times 0.45 \times 10 = 94$ ksf
Total end bearing capacity $= q \times$ area
Total end bearing capacity $= q \times \pi \times (d^2)/4 = 74$ kips
$= 37$ tons (329 kN)

NAVFAC DM 7.2

$q = \sigma'_t \times N_q$
$q =$ end bearing capacity of the pile (units same as σ'_v)
$\sigma'_t =$ effective stress at pile tip

4.1.1   Bearing Capacity Factor (N_q)

Table 4.1   Friction angle vs. Nq

ϕ	26	28	30	31	32	33	34	35	36	37	38	39	40
N_q (for driven piles)	10	15	21	24	29	35	42	50	62	77	86	120	145
N_q (for bored piles)	5	8	10	12	14	17	21	25	30	38	43	60	72

(*Source*: NAVFAC DM 7.2)

Table 4.1 shows that the N_q value is lower in bored piles; this is expected. During pile driving, the soil just below the pile tip will be compacted. Hence, it is reasonable to assume a higher N_q value for driven piles.

If water jetting is used, ϕ should be limited to 28°. This is because water jets tend to loosen the soil. Hence, higher friction angle values are not warranted.

4.1.2   Kulhawy (1984)

Kulhawy reported the following values for the end bearing capacity using pile load test data.

Table 4.2 End bearing capacity in sandy soils

Depth		Saturated Loose Sand		Dry Loose Sand		Saturated Very Dense Sand		Dry Very Dense Sand	
feet (ft)	*meters (m)*	*tsf*	*MN/m²*	*Tsf*	*MN/m²*	*tsf*	*MN/m²*	*tsf*	*MN/m²*
20	6.1	10	0.95	50	4.8	60	5.7	140	13.4
40	12.2	25	2.4	60	5.7	110	10.5	200	19.2
60	18.3	40	3.8	70	6.7	160	15.3	250	23.9
80	24.4	50	4.8	90	8.6	200	19.2	300	28.7
100	30.5	55	5.3	110	10.5	230	22.0	370	35.4

(*Source:* Kulhawy, 1984)

Design Example 4.1

Find the end bearing capacity of a 3-ft (1-m) diameter caisson using the Kulhawy values. Assume that the soil is saturated at tip level of the caisson and that the average SPT (N) value at the tip level is 10 blows/ft.

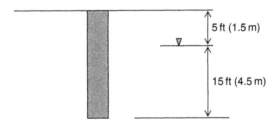

Figure 4.5 Pile diagram

Average N values at tip level = 10

Solution

Average N values of 10 blows per foot can be considered as loose sand. Hence, soil below the pile tip can be considered to be loose sand. The depth to the bottom of the pile is 20 ft.

Ultimate end bearing capacity for saturated loose sand at 20 ft = 10 tsf (0.95 MN/m²) (see Table 4.2).

Cross-sectional area of the caisson = $(\pi \times D^2/4) = 7.07$ sq ft (0. 785 m²)

Ultimate end bearing capacity = Area × 10 tons = 70.7 tons (0.74 MN)

Some of the values in the table provided by Kulhawy are very high. For instance, a caisson placed in very dense dry soil at 20 ft gives 140 tsf. (see Table 4.2)

Use of other methods is recommended to check the values obtained using tables provided by Kulhawy.

References

American Petroleum Institute, *Recommended Practice for Planning, Designing and Constructing Fixed Offshore Platforms*, API RP2A, 15th ed., 1984.

Kulhawy, F.H., *Limiting Tip and Side Resistance, Analysis and Design of Pile Foundations*, ASCE, New York, 1984.

Martin, E., et al., "Concrete Pile Design in Tidewater," *ASCE J. of Geotechnical Eng.* 113, No. 6, 1987.

NAVFAC DM 7.2, *Foundation and Earth Structures*, U.S. Department of the Navy, 1984.

4.2   Equations for Skin Friction in Sandy Soils

Numerous techniques have been proposed to compute the skin friction in piles in sandy soils. These different methodologies are briefly discussed in this chapter.

McClelland (1974) (Driven Piles)

McClelland (1974) suggested the following equation:

$$S = \beta \cdot \sigma'_v A_p$$

S = skin friction
σ'_v = effective stress at midpoint of the pile
S = total skin friction

σ'_v changes along the length of the pile. Hence, σ'_v at the midpoint of the pile should be taken.

A_p = perimeter surface area of the pile

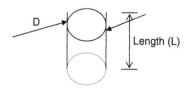

Figure 4.6   Pile parameters

Perimeter surface area of a circular pile $(A_p) = \pi \cdot D \cdot L$

$\beta = 0.15$ to 0.35 for compression

$\beta = 0.10$ to 0.25 for tension (for uplift piles)

Meyerhof (1976) (Driven Piles)

Meyerhof (1976) suggested the following equations for driven piles.

$$S = \beta \cdot \sigma'_v A_p$$

S = skin friction

σ'_v = effective stress at the midpoint of the pile

A_p = perimeter surface area of the pile

$\beta = 0.44$ for $\phi' = 28°$

$\beta = 0.75$ for $\phi' = 35°$

$\beta = 1.2$ for $\phi' = 37°$

Meyerhof (1976) (Bored Piles)

Meyerhof (1976) suggested the following equations for bored piles.

$$S = \beta \cdot \sigma'_v A_p$$

S = skin friction of the pile

σ'_v = effective stress at the midpoint of the pile

A_p = perimeter surface area of the pile

$\beta = 0.10$ for $\phi = 33°$

$\beta = 0.20$ for $\phi = 35°$

$\beta = 0.35$ for $\phi = 37°$

Kraft and Lyons (1974)

$$S = \beta \cdot \sigma'_v A_p$$

S = skin friction of the pile

σ'_v = effective stress at the midpoint of the pile

$$\beta = C \cdot \tan(\phi - 5)$$

$C = 0.7$ for compression

$C = 0.5$ for tension (uplift piles)

NAVFAC DM 7.2

$$S = K \cdot \sigma'_v \times \tan \delta \times A_p$$

S = skin friction of the pile
σ'_v = effective stress at the midpoint of the pile
K = lateral earth pressure coefficient
δ = pile skin friction angle

Table 4.3 Pile skin friction angle (δ)

Pile Type	δ
Steel piles	$20°$
Timber piles	$^3/_4\,\phi$
Concrete piles	$^3/_4\,\phi$

(*Source:* NAVFAC DM 7.2)

δ is the skin friction angle between pile material and surrounding sandy soils. Usually, smooth surfaces tend to have less skin friction compared to rough surfaces.

Table 4.4 Lateral earth pressure coefficient (K)

Pile Type	K (piles under compression)	K (piles under tension— uplift piles)
Driven H-piles	0.5–1.0	0.3–0.5
Driven displacement piles (round and square)	1.0–1.5	0.6–1.0
Driven displacement tapered piles	1.5–2.0	1.0–1.3
Driven jetted piles	0.4–0.9	0.3–0.6
Bored piles (less than 24″ diameter)	0.7	0.4

(*Source:* NAVFAC DM 7.2)

The lateral earth pressure coefficient is less in uplift piles than in regular piles. Tapered piles tend to have the highest K value.

Average K Method

Earth pressure coefficient K can be averaged from K_a, K_p, and K_0.

$$K = (K_0 + K_a + K_p)/3$$

Equations for K_0, K_a, and K_p are

$K_0 = 1 - \sin \varphi$	(earth pressure coefficient at rest)
$K_a = 1 - \tan^2(45 - \varphi/2)$	(active earth pressure coefficient)
$K_p = 1 + \tan^2(45 + \varphi/2)$	(passive earth pressure coefficient)

References

Dennis N.D., and Olson R.E., "Axial Capacity of Steel Pipe Piles in Sand," Proc. ASCE Conf. on Geotechnical Practice in Offshore Eng., Austin, TX, 1983.

Kraft, L.M., and Lyons, C.G., "State of the Art: Ultimate Axial Capacity of Grouted Piles," Proc. 6th Annual OTC, Houston paper OTC 2081, 487–503, 1990.

McClelland, B., "Design of Deep Penetration Piles for Ocean Structures," *ASCE J. of Geotechnical Eng.*, GT7, 705–747, 1974.

Meyerhof, G.G., "Bearing Capacity and Settlement of Pile Foundations," *ASCE J. of Geotechnical Eng.*, GT3, 195–228, 1976.

NAVFAC DM 7.2—*Foundation and Earth Structures*, U.S. Department of the Navy, 1984.

Olson R.E., "Axial Load Capacity of Steel Pipe Piles in Sand," Proc. Offshore Technology Conference, Houston, TX, 1990.

Stas, C.V., and Kulhawy, F.H., "Critical Evaluation of Design Methods for Foundations Under Axial Uplift and Compression Loading," Report for EPRI. No. EL. 3771, Cornell University, 1986.

Terzaghi, K., Peck, R.B., and Mesri, G., *Soil Mechanics and Foundation Engineering*, John Wiley and Sons, New York, 1996.

4.2.1 Design Examples

Design Example 4.2: Single Pile in Uniform Sand Layer (no groundwater present)

A 0.5-m (1.64-ft) diameter, 10-m (32.8-ft) long, round steel pipe pile is driven into a sandy soil stratum as shown in Figure 4.7. Compute the ultimate bearing capacity of the pile.

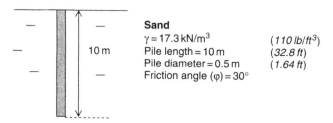

Figure 4.7   Pile in sandy soil

STEP 1: Compute the end bearing capacity

$$Q_{ultimate} = \underbrace{[\sigma'_t \times N_q \times A]}_{End\ Bearing\ Term} + \underbrace{[K \cdot \sigma'_p \times \tan \delta \times (\pi \times d) \times L]}_{Skin\ Friction\ Term}$$

End bearing term $= \sigma'_t \times N_q \times A$ ($\sigma'_t =$ effective stress at the tip of the pile)

$\sigma'_t = \gamma.$ depth to the tip of the pile $= 17.3 \times 10 = 173$ kN/m² *(3,610 psf)*

Find N_q using Table 4.1.

For a friction angle of 30, $N_q = 21$ for driven piles.

End bearing capacity $= \sigma'_t \times N_q \times A = 173 \times 21 \times (\pi \cdot d^2/4)$
$= 713.3$ kN   *(160 kips)*

STEP 2: Compute the skin friction.

Skin friction term $= K \cdot \sigma'_p \times \tan \delta \times (\pi \times d) \times L$

Obtain the K value.

From Table 4.4, for driven round piles the K value lies between 1.0 and 1.5. Hence assume $K = 1.25$.

Obtain the σ'_p (effective stress at the perimeter of the pile).

The effective stress along the perimeter of the pile varies with the depth. Hence, obtain the σ'_p value at the midpoint of the pile.

The pile is 10 m long; hence, use the effective stress at 5 m below the ground surface.

σ'_p (midpoint) $= 5 \times \gamma = 5 \times 17.3 = 86.5$ kN/m² *(1.8 ksf)*

Obtain the skin friction angle (δ).

From Table 4.3, the skin friction angle for steel piles is 20°.

Find the skin friction of the pile.

$$\text{Skin friction} = K \cdot \sigma'_p \times \tan \delta \times (\pi \times d) \times L$$
$$= 1.25 \times 86.5 \times (\tan 20) \times (\pi \times 0.5) \times 10$$
$$= 618.2 \text{ kN } \textit{(139 kips)}$$

STEP 3: Compute the ultimate bearing capacity of the pile.

$Q_{ultimate}$ = Ultimate bearing capacity of the pile
$Q_{ultimate}$ = End-bearing capacity + skin friction
$Q_{ultimate}$ = $713.3 + 618.2 = 1,331.4$ kN *(300 kips)*
Assume a factor of safety of 3.0.
Hence, allowable bearing capacity of the pile = $Q_{ultimate}$/FOS
FOS = Factor of Safety
Assume FOS = 3.0
$$= 1,331.4/3.0$$
Allowable pile capacity = 443.8 kN *(99.8 kips)*

Note: 1 kN is equal to 0.225 Kips. Hence, allowable capacity of the pile is 99.8 Kips.

Design Example 4.3: Single Pile in Uniform Sand Layer (groundwater present)

As shown in Figure 4.8, 0.5-m-diameter, 10-m-long round concrete pile is driven into a sandy soil stratum. Groundwater is located 3 m below the surface. Compute the ultimate bearing capacity of the pile.

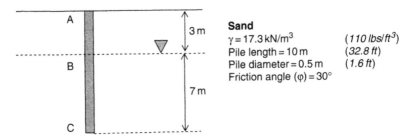

Figure 4.8   Pile in sandy soil with groundwater present

STEP 1: Compute the end bearing capacity

$$Q_{ultimate} = \underbrace{[\sigma'_t \times N_q \times A]}_{End\ Bearing\ Term} + \underbrace{[K \cdot \sigma'_v \times \tan \delta \times (\pi \times d) \times L]}_{Skin\ Friction\ Term}$$

End bearing term $= \sigma'_t \times N_q \times A$ (σ'_t = effective stress at the tip of the pile)

$\sigma'_t = 17.3 \times 3 + (17.3 - \gamma_w) \times 7$

γ_w = density of water $= 9.8$ kN/m^3

$= 17.3 \times 3 + (17.3 - 9.8) \times 7 = 104.4$ kN/m^2 (2,180 psf)

Find N_q using Table 4.1.

For a friction angle of 30, $N_q = 21$ for driven piles.

End bearing capacity $= \sigma'_t \times N_q \times A$

$$= 104.4 \times 21 \times (\pi \cdot 0.5^2/4) = 403.5 \text{ kN} \ (96.8 \ kips)$$

STEP 2: Compute the skin friction (A to B).

Skin friction has to be computed in two parts. First find the skin friction from A to B and then find the skin friction from B to C.

Skin friction term $= K \cdot \sigma'_v \times \tan \delta \times (\pi \times d) \times L$

Obtain the K value.
From Table 4.4, for driven round piles the K value lies between 1.0 and 1.5. Hence, assume $K = 1.25$.

Obtain the σ'_v (effective stress at the perimeter of the pile).
The effective stress along the perimeter of the pile varies with the depth. Hence, obtain the σ'_p value at the midpoint of the pile from point A to B. The length of pile section from A to B is 3 m. Hence, find the effective stress at 1.5 m below the ground surface.

Obtain the skin friction angle (δ),
From Table 4.3, the skin friction angle for steel piles is 20°.

$$\text{Skin friction} = K \cdot \sigma'_v \times \tan \delta \times (\pi \times d) \times L$$

σ'_v = Effective stress at midpoint of section A to B (1.5 m below the surface).

$\sigma'_v = 1.5 \times 17.3 = 25.9$ kN/m^2 (0.54 Ksf)

$K = 1.25$, $\delta = 20°$

Skin friction (A to B) $= 1.25 \times (25.9) \times (\tan 20) \times (\pi \times 0.5) \times 3$

$$= 55.5 \text{ kN} \ (12.5 \ kips)$$

Find the skin friction of the pile from B to C.

$$\text{Skin friction} = K \cdot \sigma'_v \times \tan \delta \times (\pi \times d) \times L$$

σ'_v = Effective stress at midpoint of section B to C = $3 \times 17.3 +$
(17.3 − 9.8) × 3.5
σ'_v = Effective stress at midpoint of section B to C = 78.2 kN/m^2
(*1,633 psf*)
K = 1.25 and δ = 20°
Skin friction (B to C) = 1.25 × (78.2) × tan (20) × (π × 0.5) × 7
= 391 kN   (*87.9 kips*)

STEP 3: Compute the ultimate bearing capacity of the pile.

$Q_{ultimate}$ = ultimate bearing capacity of the pile
$Q_{ultimate}$ = end bearing capacity + skin friction
$Q_{ultimate}$ = 403.5 + 55.5 + 391 = 850 kN
Assume a factor of safety of 3.0.
Hence, allowable bearing capacity of the pile = $Q_{ultimate}$/FOS
= 850.1/3.0 = 283 kN
Allowable pile capacity = 283 kN   (*63.6 kips*)

Note: 1 kN is equal to 0.225 kips.

Design Example—4.4 (multiple sand layers with no groundwater present)

As shown in figure 4.9, a 0.5-m-diameter, 12-m-long round concrete pile is driven into a sandy soil stratum. Compute the ultimate bearing capacity of the pile.

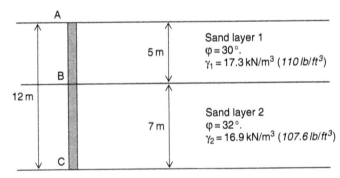

Figure 4.9   Pile in multiple sand layers with no groundwater present

Solution

STEP 1: Compute the end bearing capacity:

$$Q_{ultimate} = \underbrace{[\sigma'_t \times N_q \times A]}_{End\ Bearing\ Term} + \underbrace{[K \cdot \sigma'_v \times \tan \delta \times A_p]}_{Skin\ Friction\ Term}$$

End bearing term $= \sigma'_t \times N_q \times A$
($\sigma'_t =$ Effective stress at the tip of the pile)
$\sigma'_t = \gamma_1 \cdot 5 + \gamma_2 \cdot 7$
$\sigma'_t = 17.3 \times 5 + 16.9 \times 7 = 204.8$ kN/m^2 *(4.28 ksf)*
Find N_q using Table 4.1. Use the friction angle of soil where the pile tip rests.
For a friction angle of 32, $N_q = 29$ for driven piles. The Diameter of the pile is 0.5 m.
The φ value of the bottom sand layer is used to find N_q, since the tip of the pile lies on the bottom sand layer.
End bearing capacity $= \sigma'_t \times N_q \times A$
$$= 204.8 \times 29 \times (\pi \cdot d^2/4) = 1{,}166.2 \text{ kN } (262.2 \text{ kips})$$

STEP 2: Compute of the skin friction.

The skin friction of the pile needs to be done in two parts.
Skin friction of the pile portion in sand layer 1 (A to B)
Skin friction of the pile portion in sand layer 2 (B to C)

Compute the skin friction of the pile portion in sand layer 1 (A to B).

$$\text{Skin friction term} = K \cdot \sigma'_p \times \tan \delta \times A_p$$
$$A_p = (\pi \times d) \times L$$

Obtain the K value.
From Table 4.4, for driven round piles the K value lies between 1.0 and 1.5.
Hence, assume $K = 1.25$.

Obtain the σ'_v (average effective stress at the perimeter of the pile).
Obtain the σ'_v value at the midpoint of the pile in sand layer 1.

$$\sigma'_v \text{ (midpoint)} = 2.5 \times (\gamma_1) = 2.5 \times 17.3 \text{ kN/m}^2$$
$$\sigma'_v \text{ (midpoint)} = 43.3 \text{ kN/m}^2 \quad (0.9 \text{ ksf})$$

Obtain the skin friction angle (δ).
From Table 4.3, the skin friction angle (δ) for concrete piles is $^3/_4 \varphi$.

$$\delta = {}^3/_4 \times 30° = 22.5° \text{ (Friction angle of layer 1 is 30°.)}$$

Skin friction in sand layer 1 $= K \cdot \sigma'_v \times \tan \delta \times (\pi \times d) \times L$
$$= 1.25 \times 43.3 \times (\tan 22.5) \times (\pi \times 0.5) \times 5$$
$$= 176.1 \text{ kN} \quad (39.6 \text{ kips})$$

STEP 3: Find the skin friction of the pile portion in sand layer 2 (B to C).

Skin friction term $= K \cdot \sigma'_v \times \tan \delta \times (\pi \times d) \times L$

Obtain the σ'_v (effective stress at the perimeter of the pile).

Obtain the σ'_v value at the midpoint of the pile in sand layer 2.

$$\sigma'_v \text{ (mid point)} = (5 \times \gamma_1) + (3.5 \times \gamma_2)$$
$$\sigma'_v \text{ (mid point)} = 5 \times 17.3 + 3.5 \times 16.9 \text{ kN/m}^2$$
$$= 145.7 \text{ kN/m}^2 \ (3043 \text{ psf})$$

Obtain the skin friction angle (δ).
From Table 4.3, the skin friction angle (δ) for concrete piles is $^3/_4 \varphi$.

$$\delta = {}^3/_4 \times 32° = 24° \text{ (Friction angle of layer 2 is 32°.)}$$

Skin friction in sand layer 2 $= K \cdot \sigma'_p \times \tan \delta \times (\pi \times d) \times L$
$$= 1.25 \times 145.7 \times (\tan 24) \times (\pi \times 0.5) \times 7$$
$$= 891.6 \text{ kN} \quad (200.4 \text{ kips})$$

$$P_{ultimate} = \text{end bearing capacity} + \text{skin friction in layer 1}$$
$$+ \text{skin friction in layer 2}$$

$$\text{End bearing capacity} = 1{,}166.2 \text{ kN}$$
$$\text{Skin friction in sand layer 1} = 176.1 \text{ kN}$$
$$\text{Skin friction in sand layer 2} = 891.6 \text{ kN}$$

$$P_{\text{ultimate}} = 2{,}233.9 \text{ kN} \quad \textit{(502.2 kips)}$$

One can see that the bulk of the pile capacity comes from the end
  bearing. Next, is the skin friction in layer 2 (bottom layer). Skin
  friction in the top layer is very small. One reason is that the effective
  stress acting on the perimeter of the pile is very low in the top layer.
  This is because effective stress is directly related to depth.
Hence, allowable bearing capacity of the pile $= P_{\text{ultimate}}/\text{FOS}$
FOS $=$ Factor of safety $= 3.0$
     $= 2{,}233.9/3.0 \text{ kN}$
Allowable pile capacity $= 744.6 \text{ kN} \quad \textit{(167.4 kips)}$

Design Example—4.5 (multiple sand layers with groundwater present)

As shown in Figure 4.10, a 0.5-m-diameter, 15-m-long round concrete
pile is driven into a sandy soil stratum. Groundwater is 3 m below the
surface. Compute the ultimate bearing capacity of the pile.

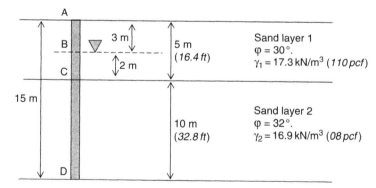

Figure 4.10   Pile in multiple sand layers with groundwater present

Solution

STEP 1: Compute the end bearing capacity

$$Q_{ultimate} = \underbrace{[\sigma'_t \times N_q \times A]}_{\text{End Bearing Term}} + \underbrace{[K \cdot \sigma'_v \times \tan \delta \times A_p]}_{\text{Skin Friction Term}}$$

End bearing term $= \sigma'_t \times N_q \times A$
($\sigma'_t =$ effective stress at the tip of the pile)
$\sigma'_t = \gamma_1 \cdot 3 + (\gamma_1 - \gamma_w) \cdot 2 + (\gamma_2 - \gamma_w)10$
$\sigma'_t = 17.3 \times 3 + (17.3 - 9.8) \times 2 + (16.9 - 9.8) \times 10$
$\quad\ = 137.9 \text{ kN/m}^2$ *(2,880 psf)*
Find N_q using Table 4.1.
For a friction angle of 32, $N_q = 29$ for driven piles.
The ϕ value of the bottom sand layer is used to find N_q, since the tip of
the pile lies on the bottom sand layer.
End bearing capacity $= \sigma'_t \times N_q \times A$
$$= 137.9 \times 29 \times (\pi \cdot d^2/4) = 785.2 \text{ kN} \quad (176.5 \text{ kips})$$

STEP 2: Compute the skin friction.

The skin friction of the pile needs to be done in three parts.
Skin friction of the pile portion in sand layer 1 above groundwater
(A to B)
Skin friction of the pile portion in sand layer 1 below groundwater
(B to C)
Skin friction of the pile portion in sand layer 2 below groundwater
(C to D)

STEP 3: Compute the skin friction of the pile portion in *sand layer 1*
above groundwater (A to B).

$$\text{Skin friction term} = K \cdot \sigma'_v \times \tan \delta \times (\pi \times d) \times L$$

Obtain the K value.
From Table 4.4, for driven round piles, the K value lies between 1.0 and
1.5.
Hence, assume $K = 1.25$.

Obtain the σ'_v (effective stress at the perimeter of the pile).
Obtain the σ'_v value at the midpoint of the pile in sand layer 1, above groundwater (A to B).

$$\sigma'_v \text{ (midpoint)} = 1.5 \times (\gamma_1) = 1.5 \times 17.3 \text{ kN/m}^2$$
$$\sigma'_v \text{ (midpoint)} = 26 \text{ kN/m}^2 \quad (543 \text{ psf})$$

Obtain the skin friction angle (δ).
From Table 4.3, the skin friction angle (δ) for concrete piles is $^3/_4 \, \varphi$.

$$\delta = {}^3/_4 \times 30° = 22.5° \quad \text{(Friction angle of layer 1 is } 30°)$$

Skin friction in sand layer 1 (A to B) $= K \cdot \sigma'_v \times \tan \delta \times (\pi \times d) \times L$

$$= 1.25 \times (26) \times (\tan 22.5)$$
$$\times (\pi \times 0.5) \times 3$$
$$= 63.4 \text{ kN} \quad (14.3 \text{ kips})$$

STEP 4: Find the skin friction of the pile portion in *sand layer 1* below groundwater (B to C).

$$\text{Skin friction term} = K \cdot \sigma'_v \times \tan \delta \times (\pi \times d) \times L$$

Obtain the σ'_v (effective stress at the perimeter of the pile).

$$\sigma'_v \text{ (midpoint)} = 3 \times \gamma_1 + 1.0 \, (\gamma_1 - \gamma_w)$$
$$\sigma'_v \text{ (midpoint)} = 3 \times 17.3 + 1.0 \times (17.3 - 9.8) \text{ kN/m}^2$$
$$= 59.4 \text{ kN/m}^2 \quad (1{,}241 \text{ psf})$$

Skin friction in sand layer 1 (B to C) $= K \cdot \sigma'_p \times \tan \delta \times (\pi \times d) \times L$

$$= 1.25 \times 59.4 \times (\tan 22.5)$$
$$\times (\pi \times 0.5) \times 2$$
$$= 96.6 \text{ kN} \quad (21.7 \text{ kips})$$

STEP 5: Find the skin friction in *sand layer 2* below groundwater (C to D).

$$\text{Skin friction term} = K \cdot \sigma'_v \times \tan \delta \times (\pi \times d) \times L$$

Obtain the σ'_p (effective stress at the midpoint of the pile).

$$\sigma'_v \text{ (midpoint)} = 3 \times \gamma_1 + 2(\gamma_1 - \gamma_w) + 5 \times (\gamma_2 - \gamma_w)$$
$$\sigma'_v \text{ (midpoint)} = 3 \times 17.3 + 2.0 \times (17.3 - 9.8) + 5 \times (16.9 - 9.8) \text{ kN/m}^2$$
$$= 102.4 \text{ kN/m}^2 \quad (2.14 \text{ ksf})$$

From Table 4.3, the skin friction angle (δ) for concrete piles is $^3/_4\,\varphi$.

$$\delta = {}^3/_4 \times 32° = 24° \quad \text{(Friction angle of layer 2 is 32°.)}$$

Skin friction in sand layer 2 (C to D) $= K \cdot \sigma'_v \times \tan \delta \times (\pi \times d) \times L$
$$= 1.25 \times 102.4 \times (\tan 24)$$
$$\times (\pi \times 0.5) \times 10$$
$$= 895.2 \text{ kN} \quad (201.2 \text{ kips})$$

$P_{ultimate}$ = end bearing capacity + skin friction in layer 1 (above GW)
    + skin friction in layer 1 (below GW)
    + skin friction in layer 2 (below GW)

End bearing capacity $= 785.2$ kN
Skin friction in layer 1 (above GW) (A to B) $= 63.4$ kN
Skin friction in layer 1 (below GW) (B to C) $= 96.6$ kN
Skin friction in layer 2 (below GW) (C to D) $= 895.2$ kN
Total $= 1,840$ kN

$$P_{ultimate} = 1,840.4 \text{ kN} \quad (413.7 \text{ kips})$$

In this case, the bulk of the pile capacity comes from the end bearing and the skin friction in the bottom layer.

Hence, allowable bearing capacity of the pile $= P_{ultimate}/\text{FOS}$
FOS = Factor of safety = 3.0
$$= 1,840.4/3.0$$
Allowable pile capacity $= 613.5$ kN (137.9 kips)

4.3 Pile Design Using the Meyerhof Equation (Correlation with SPT (N)

4.3.1 End Bearing Capacity

Meyerhof proposed the following equation based on SPT (N) value to compute the ultimate end bearing capacity of driven piles. The Meyerhof equation was adopted by DM 7.2 as an alternative method to static analysis.

$$q_{ult} = 0.4 \ C_N \times N \times D/B$$

q_{ult} = ultimate point resistance of driven piles (tsf)
  N = standard penetration resistance (blows/ft) near pile tip
C_N = 0.77 log 20/p
  p = effective overburden stress at pile tip (tsf)

Effective stress "p" should be more than 500 psf. It is very rare for effective overburden stress at pile tip to be less than 500 psf.

D = depth driven into granular (sandy) bearing stratum (ft)
B = width or diameter of the pile (ft)
q_1 = limiting point resistance (tsf), equal to 4N for sand and 3N for silt
q_{ult} should not exceed 4N (tsf) for sand and 3N (tsf) for silt.

4.4   Modified Meyerhof Equation

Meyerhof developed the above equation using many available load test data and obtaining average N values. Pile tip resistance is a function of the friction angle. For a given SPT (N) value, different friction angles are obtained for different soils.

For a given SPT (N) value, friction angle for coarse sand is 7 to 8% higher compared to medium sand. At the same time, for a given SPT (N) value, friction angle is 7 to 8% lower in fine sand compared to medium sand. For this reason, the following modified equations can be proposed.

$$q_{ult} = 0.45\,C_N \times N\;D/B \text{ tsf (coarse sand)}$$
$$q_{ult} = 0.4\,C_N \times N\;D/B \text{ tsf (medium sand)}$$
$$q_{ult} = 0.35\,C_N \times N\;D/B \text{ tsf (fine sand)}$$

Design Example 4.6

Find the tip resistance of the 2-ft (0.609-m) diameter pile shown using the Meyerhof equation. SPT (N) value at pile tip is 25 blows per foot.

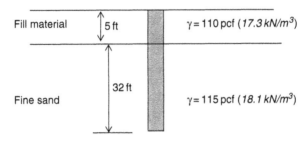

Figure 4.11   Pile in fill material underlain by fine sand

Solution

STEP 1: Ultimate point resistance for driven piles for fine sand

$q_{ult} = 0.35 \ C_N \times N \times D/B$ tsf (fine sand)
$C_N = 0.77 \log 20/p$
p = effective overburden stress at pile tip (tsf)
$p = 5 \times 110 + 32 \times 115 = 4{,}230$ psf $= 2.11$ tsf   *(0.202 MPa)*
$C_N = 0.77 \log [20/2.11] = 0.751$
D = depth driven into bearing stratum = 32 ft   *(9.75 m)*
Fill material is not considered to be a bearing stratum.
$B = 2$ ft (width or diameter of the pile) *(0.61 m)*
$q_{ult} = 0.35 \ C_N \times N \times D/B$ (fine sand)
$q_{ult} = 0.35 \times 0.751 \times 25 \times 32/2 = 105$ tsf   *(10.1 MPa)*
Maximum allowable point resistance $= 4 \ N$ for sandy soils
$4 \times N = 4 \times 25 = 100$ tsf
Hence, $q_{ult} = 100$ tsf   *(9.58 MPa)*
Allowable point bearing capacity $= 100/FOS$
FOS = Factor of safety.
Assume a factor of safety of 3.0
Hence, $q_{allowable} = 33.3$ tsf   *(3.2 MPa)*
Total allowable point bearing capacity $= q_{allowable} \times$ Tip area
$\qquad\qquad\qquad\qquad = q_{allowable} \times \pi \times (2^2)/4$
$\qquad\qquad\qquad\qquad = 105$ tons *(934 kN)*

Meyerhof Equations for Skin Friction

Meyerhof proposes the following equation for skin friction:

$$f = N/50 \text{ tsf}$$

f = unit skin friction (tsf)
N = average SPT (N) value along the pile

Note: As per Meyerhof, unit skin friction "f" should not exceed 1 tsf.

The following modified equations can be proposed to account for soil gradation.

f = N/46 tsf   coarse sand
f = N/50 tsf   medium sand
f = N/54 tsf   fine sand

Design Example 4.7

Find the skin friction of the 2-ft-diameter pile shown using the Meyerhof equation. The average SPT (N) value along the shaft is 15 blows per foot.

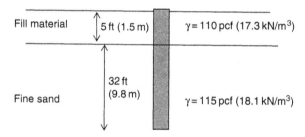

Figure 4.12   Skin friction of a pile in fill material underlain by fine sand

Solution

STEP 1: Ignore the skin friction in fill material.

For fine sand:
  Unit skin friction; f = N/54 tsf
  Unit skin friction; f = 15/54 tsf = 0.28 tsf   *(26.8 kN/m²)*
Total skin friction = unit skin friction × perimeter surface area
Total skin friction = 0.28 × π × D × L
Total skin friction = 0.28 × π × 2 × 32 = 56 tons
Allowable skin friction = 56/FOS
Assume a (FOS) factor of safety of 3.0
Allowable skin friction = 56/3.0 = 18.7 tons

4.5   Parameters that Affect the End Bearing Capacity

The following parameters affect the end bearing capacity.

- Effective stress at pile tip
- Friction angle at pile tip and below (ϕ')
- The dilation angle of soil (ψ)
- Shear modulus (G)
- Poisson's ratio (v)

Most of these parameters have been bundled into the bearing capacity factor (N_q). It is known that the friction angle decreases with the depth. Hence N_q, which is a function of the friction angle, also would reduce with depth. Variation of other parameters with depth has not been researched thoroughly.

Increase of end bearing capacity do not increase at the same rate with increasing depth.

The Figure 4.13 is an attempt to formulate the end bearing capacity of a pile with regard to relative density (RD) and effective stress.

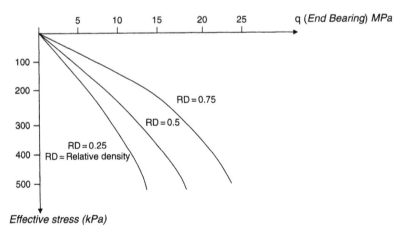

Figure 4.13   Relative density of soil. (*Source*: Randolph et al., 1994)

RD = Relative density (see the section "SPT and Friction Angle" to obtain the relative density value using SPT (N) values).

- As per Figure 4.13, end bearing capacity tapers down with increasing effective stress.

References

Kulhawy, F.H., et al., *Transmission Line Structure Foundations for Uplift-Compression Loading*, Report EL. 2870, Electric Power Research Institute, Palo Alto, 1983.
Randolph, M.F., Dolwin, J., and Beck, R., "Design of Driven Piles in Sand," *Geotechnique* 44, No. 3, 427–448, 1994.

4.6   Critical Depth for Skin Friction (Sandy Soils)

Vertical effective stress (σ') increases with depth. Hence, skin friction should increase with depth indefinitely. In reality, skin friction will not increase with depth indefinitely.

It was once believed that skin friction would become a constant at a certain depth. This depth was named critical depth.

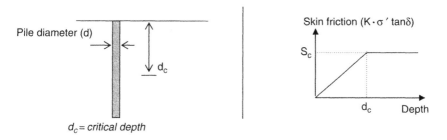

Figure 4.14   Critical depth

- As shown in Figure 4.14, skin friction was assumed to increase till the critical depth and then maintain a constant value.

  d_c = critical depth
  S_c = skin friction at critical depth ($K \cdot \sigma'_c \cdot \tan \delta$)
  σ'_c = effective stress at critical depth

The following approximations were assumed for the critical depth.

- Critical depth for loose sand $= 10 \, d$ (d is the pile diameter or the width.)

- Critical depth for medium dense sand $= 15 \, d$

- Critical depth for dense sand $= 20 \, d$

This theory does not explain recent precise pile load test data. According to recent experiments, skin friction will not become a constant abruptly as was once believed.

Experimental Evidence for Critical Depth

Figure 4.15 shows the typical variation of skin friction with depth in a pile.

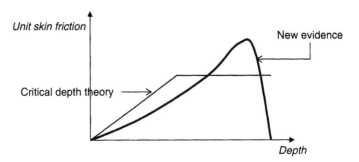

Figure 4.15   Variation of skin friction. (*Source*: Randolph et al., 1994)

- As one can see, experimental data does not support the old theory with a constant skin friction below the critical depth.
- Skin friction tends to increase with depth and just above the tip of the pile to attain its maximum value. Skin friction would drop rapidly after that.
- Skin friction does not increase linearly with depth as was once believed.
- No satisfactory theory exists at the present to explain the field data.
- Due to lack of a better theory, engineers are still using critical depth theory of the past.

4.6.1   Reasons for Limiting Skin Friction

The following reasons have been offered to explain why skin friction does not increase with depth indefinitely, as suggested by the skin friction equation.

$$\text{Unit Skin Friction} = K \cdot \sigma' \tan \delta; \quad \sigma' = \gamma \cdot d$$

1. The above K value is a function of the soil friction angle (ϕ'). Friction angle tends to decrease with depth. Hence, K value decreases with depth (*Kulhawy 1983*).

2. The above skin friction equation does not hold true at high stress levels due to readjustment of sand particles.

3. Reduction of local shaft friction with increasing pile depth. (See Figure 4.16) (*Randolph et al. 1994*).

Assume that a pile was driven to a depth of 10 ft and unit skin friction was measured at a depth of 5 ft. Then assume that the pile was driven further to a depth of 15 ft and unit skin friction was measured at the same depth of 5 ft. It has been reported that unit skin friction at 5 ft is less in the second case.

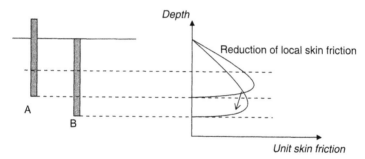

Figure 4.16   Variation of skin friction in relation to depth

According to Figure 4.16, local skin friction decreases when the pile is driven further into the ground.

As per NAVFAC DM 7.2, maximum value of skin friction and end bearing capacity is achieved after 20 diameters *within the bearing zone*.

4.7   Critical Depth for End Bearing Capacity (Sandy Soils)

• Pile end bearing capacity in sandy soils is related to effective stress. Experimental data indicates that end bearing capacity does not increase with depth indefinitely. Due to lack of a valid theory, engineers use the same critical depth concept adopted for skin friction for end bearing capacity as well.

d_c = critical depth

Figure 4.17   Critical depth for end bearing capacity

- As shown in Figure 4.17, the end bearing capacity was assumed to increase till the critical depth.

  d_c = Critical depth; Q_c = end bearing at critical depth
  σ'_c = effective stress at critical depth

The following approximations were assumed for the critical depth.

- Critical depth for loose sand = 10 d (d is the pile diameter or the width.)
- Critical depth for medium dense sand = 15 d
- Critical depth for dense sand = 20 d

  Is critical depth for end bearing the same as the critical depth for skin friction?
  Since the critical depth concept is a gross approximation that cannot be supported by experimental evidence, the question is irrelevant. It is clear that there is a connection between end bearing capacity and skin friction since the same soil properties act in both cases such as effective stress, friction angle, and relative density. On the other hand, two processes are vastly different in nature.

Critical Depth Example

Design Example 4.8: Find the skin friction and end bearing capacity of the pile shown. Assume that critical depth is achieved at 20 ft into the bearing layer (Ref. DM 7.2). Pile diameter is 1 ft, and other soil parameters are as shown in the figure.

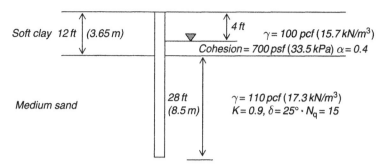

Figure 4.18   Pile in soft clay underlain by medium sand

Solution

The skin friction is calculated in the overburden soil. In this case, skin friction is calculated in the soft clay. Then the skin friction is calculated in the bearing layer (medium sand) assuming the skin friction attains a limiting value after 20 diameters (critical depth).

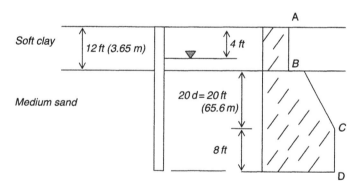

Figure 4.19   Pile skin friction diagram

STEP 1: Find the skin friction from A to B.

$$\text{Skin friction in soft clay} = \alpha \times c \times \text{perimeter surface area}$$
$$= 0.4 \times 700 \times \pi \times d \times L$$
$$= 0.4 \times 700 \times \pi \times 1 \times 12 = 10{,}560 \,\text{lbs}$$
$$= (46.9 \,\text{kN})$$

STEP 2: Find the skin friction from B to C.

Skin friction in sandy soils $= S = K \cdot \sigma'_v \times \tan \delta \times A_p$
$S =$ skin friction of the pile
$\sigma'_v =$ average effective stress along the pile shaft
Average effective stress along pile shaft from B to C $= (\sigma_B + \sigma_C)/2$
$\sigma_B =$ effective stress at B
$\sigma_C =$ effective stress at C

To obtain the average effective stress from B to C, find the effective stresses at B and C and obtain the average of those two values.

$$\sigma_B = 100 \times 4 + (100 - 62.4) \times 8 = 700.8\, \text{lb/ft}^2 \quad (33.6\, kPa)$$
$$\sigma_C = 100 \times 4 + (100 - 62.4) \times 8 + (110 - 62.4) \times 20$$
$$= 1{,}452.8\, \text{lb/ft}^2 \quad (69.5\, kPa)$$

Average effective stress along pile shaft from B to C $= (\sigma_B + \sigma_C)/2$

$$= (700.8 + 1452.8)/2 = 1076.8\, \text{lb/ft}^2.$$

Skin friction from B to C $= K \cdot \sigma'_v \times \tan \delta \times A_p$
$$= 0.9 \times 1076.8 \times \tan(25°)$$
$$\times (\pi \times 1 \times 20) = 28{,}407\, \text{lbs}$$

STEP 3: Find the skin friction from C to D.

Skin friction reaches a constant value at point C, 20 diameters into the
   bearing layer.
Skin friction at point C $= K \cdot \sigma'_v \times \tan \delta \times A_p$
σ'_v at point C $= 100 \times 4 + (100 - 62.4) \times 8 + (110 - 62.4) \times 20$
$$= 1{,}452.8\, \text{lb/ft}^2$$
Unit skin friction at point C $= 0.9 \times 1{,}452.8 \times \tan 25°$
$$= 609.7\, \text{lb/ft}^2 \quad (29\, kPa)$$
Unit skin friction is constant from C to D. This is because skin friction
   does not increase after the critical depth.

Skin friction from C to D $= 609.7 \times$ surface perimeter area
$$= 609.7 \times (\pi \times 1 \times 8)\, \text{lbs} = 15{,}323.4\, \text{lbs}$$
$$= (68.2\, kN)$$

Summary

Skin friction in soft clay (A to B) $= 10,560$ lbs
  Skin friction in sand (B to C) $= 28,407$ lbs
  Skin friction in sand (C to D) $= 15,323$ lbs
                    Total $= 54,290$ lbs (241 kN)

STEP 4: Compute the end bearing capacity.

End bearing capacity also reaches a constant value below the critical depth.

$$\text{End bearing capacity} = q \times N_q \times A$$

$q =$ effective stress at pile tip
$N_q =$ bearing capacity factor (given to be 15)
$A =$ cross-sectional area of the pile

If the pile tip is below the critical depth, q should be taken at critical depth. In this example, pile tip is below the critical depth, which is 20 diameters into the bearing layer. Hence, q is equal to the effective stress at critical depth (point C).

Effective stress at point C $= 100 \times 4 + (100 - 62.4) \times 8 + (110 - 62.4)$
$\times 20 = 1,452.8 \, \text{lb/ft}^2$
End bearing capacity $= q \times N_q \times A = 1,452.8 \times 15 \times (\pi \times d^2/4) \, \text{lbs}$
$= 17,115$ lbs
Total ultimate capacity of the pile $=$ total skin friction $+$ end bearing
$= 54,290 + 17,115 = 71,405$ lbs
$= 35.7$ tons *(317.6 kN)*

References

Kulhawy, F.H., et al., *Transmission Line Structure Foundations for Uplift-Compression Loading*, Report EL. 2870, Electric Power Research Institute, Palo Alto, 1983.

NAVFAC DM 7.2, *Foundation and Earth Structures*, U.S. Department of the Navy, 1984.

Randolph, M.F., Dolwin, J., and Beck, R., "Design of Driven Piles in Sand," *Geotechnique* 44, No. 3, 427–448, 1994.

5

Pile Design in Clay Soils

Two types of forces act on piles.

1. End bearing acts on the bottom of the pile.
2. Skin friction acts on the sides of the pile.

To compute the total load that can be applied to a pile, one needs to compute the end bearing and the skin friction acting on sides of the pile.

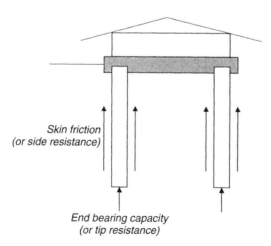

Figure 5.1 Skin friction in piles

A modified Terzaghi bearing capacity equation is used to find the pile capacity

$$P_u = Q_u + S_u$$
$$P_u = 9 \cdot c \cdot A_c + \alpha \cdot c \cdot A_p$$

P_u = ultimate pile capacity
Q_u = ultimate end bearing capacity
S_u = ultimate skin friction
c = cohesion of the soil
A_c = cross-sectional area of the pile
A_p = perimeter surface area of the pile
α = adhesion factor between pile and soil

As per the above equation, clay soils with a higher cohesion would have a higher end bearing capacity. The end bearing capacity of piles in clayey soils is usually taken to be $9 \cdot c$.

Skin Friction

Assume a block is placed on a clay surface. Now if a force is applied to move the block, the adhesion between the block and the clay will resist the movement. If the cohesion of clay is "c," the force due to adhesion will be $\alpha \cdot c$. (α is the adhesion coefficient.)

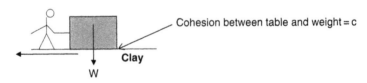

Figure 5.2   Friction acting on a solid block

α = adhesion coefficient depends on pile material and clay type
c = cohesion

Highly plastic clays would have a higher adhesion coefficient. Typically, it is assumed that adhesion is not dependent on the weight of the block. This is not strictly true as explained later. Now let's look at a pile outer surface.

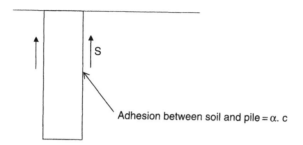

Figure 5.3   Development of skin friction

Ultimate skin friction $= S_u = \alpha \times c \cdot A_p$

It is usually assumed that ultimate skin friction is independent of the effective stress and depth. In reality, skin friction is dependent on effective stress and cohesion of soil.

The skin friction acts on the perimeter surface of the pile.

For a circular pile, the surface area of the pile is given by $\pi \cdot D \cdot L$ ($\pi \times D$) = circumference of the pile   (L = length of the pile)

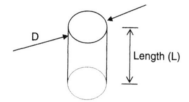

Figure 5.4   Pile dimensions

Perimeter surface area of a circular pile $(A_p) = \pi \cdot D \cdot L$

5.1   Shear Strength (Clays)

- *Tensile Failure:* When a material is subjected to tensile stress, it undergoes tensile failure.

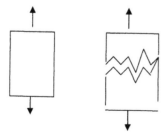

- Two figures show material failure under tension.
- For soils this type of failure is rare.
- Soils in most situations fail under compression due to shear failure.

Figure 5.5   Failure under tension

- *Shear Failure:* When a soil sample is subjected to compressive stress, the soil tends to fail under shear.

> - The figure to the left shows a soil sample being subjected to a compressive load. The soil sample is about to fail under shear.
> - Shear failure is different from tensile failure.
> - Resistance to shearing depends on soil type, water content, and drainage of water from the soil sample.

Figure 5.6   Shear failure

- *Cohesion and Friction:* Resistance to shear failure (or shear strength) is developed due to cohesion and friction (ϕ) between particles.

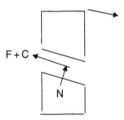

Figure 5.7   Failure plane. F = friction; C = cohesion; N = normal stress

- The top part of the soil sample is about to move to the right as shown by the top arrow.
- This movement is resisted by friction (F) and cohesion (C) existing between soil particles.
- Friction is a function of the normal stress (N).
- Pure sand have no cohesion.

5.2   Cohesion in Clay Soils

How does one measure the cohesion of clays?

Unconfined Compressive Strength Test: The easiest and most common test done to measure the cohesion is the unconfined compressive strength test (commonly known as the unconfined test). In this test, a soil sample is obtained using a Shelby tube, and a load is applied. No cell pressure is applied.

- Since there is no confining cell pressure, the test is given the adjective "unconfined."

- Since the load is applied rapidly, there is no time for the soil sample to drain. Hence, this test is an "undrained" test.

Figure 5.8 Undrained test

5.2.1 Unconfined Compressive Strength Test

- The load is applied rapidly. Hence, pore pressures will develop inside the soil sample.

- When the soil sample is under high pore pressure, friction between particles will be minimal. This is understandable since, due to water, soil particles will not have a great chance of contacting each other.

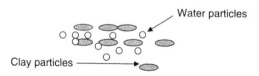

Figure 5.9 Undrained condition

Contact between clay particles is limited owing to water particles. Hence, *frictional force* due to contact between soil particles is negligible.

Electromagnetic *cohesion* between clay particles dominates.

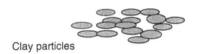

Clay particles

Figure 5.10 Drained condition

Unlike the undrained condition in which there were water particles, now during the "drained" condition contact between clay particles is high. Hence *frictional force* due to soil particle contact is high. Electromagnetic *cohesion* between clay particles also exists.

- It is obvious that the drain condition would have a higher shear strength than the undrained condition.
- If the load is applied slowly, then the soil sample will have enough time to drain. Hence, soil particles will get a chance to contact each other. This will increase the friction between particles.
- If the load is applied rapidly, then soil sample will not have enough time to drain. The contact between soil particles is limited owing to water molecules. Hence, frictional forces become negligible.

5.2.2 Parameters

The total capacity of a pile depends on many parameters.

- Skin friction
- End bearing capacity
- Negative skin friction
- Pile properties (toe area, perimeter area, pile material, pile flexibility)
- Driving process (driven, jetted down, vibrated down, or jacked down)
- Hammer type, hammer weight, and stroke for driven piles
- Loading type (cyclic and vibratory loads reduce the pile capacity)

5.3   End Bearing Capacity in Clay Soils (Different Methods)

5.3.1   Driven Piles

Skempton (1959)

The equation proposed by Skempton is widely being used to find the end bearing capacity in clay soils.

$$q = 9 \cdot Cu$$

q = end bearing capacity; C_u = cohesion of soil at the tip of the pile

Martin et al. (1987)

$$q = C \cdot N \ MN/m^2$$

C = 0.20;   N = SPT value at pile tip

5.3.2   Bored Piles

Shioi and Fukui (1982)

$$q = C \cdot N \ MN/m^2$$

C = 0.15;   N = SPT value at pile tip

5.3.3   NAVFAC DM 7.2

$$q = 9 \cdot C_u$$

q = end bearing capacity; C_u = cohesion of soil at the tip of the pile

References

Martin, R., et al., "Concrete Pile Design in Tidewater," *ASCE, J. Geotechnical Eng.* 113, No. 6, 1987.

Shioi, Y., and Fukui, J., "Application of 'N' Value to Design of Foundations in Japan," *Proc. ESOPT2,* Amsterdam 1, 159–164, 1982.

Skempton, A.W., "Cast-in-situ Bored Piles in London Clay," *Geotechnique,* 153–173, 1959.

5.4 Skin Friction in Clay Soils (Different Methods)

5.4.1 Equation Based on Undrained Shear Strength (Cohesion)
$\rightarrow f_{ult} = \alpha \times S_u$

f_{ult} = ultimate skin friction
α = skin friction coefficient
S_u = undrained shear strength or cohesion
$S_u = Q_u/2$;   (Q_u = Unconfined compressive strength)

This equation ignores effective stress effects.

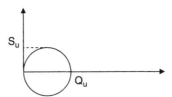

Figure 5.11   Mohr's diagram

Driven Piles

American Petroleum Institute (API) (1984)
API provides the following equation to find the skin friction in clay soils.

$$f = \alpha \cdot C_u$$

f = unit skin friction; C_u = cohesion
α = 1.0 for clays with $C_u < 25 \, kN/m^2$ *(522 psf)*
α = 0.5 for clays with $C_u > 70 \, kN/m^2$ *(1,460 psf)*

Interpolate for the α value for cohesion values between 25 kN/m^2 and 70 kN/m^2.

As per the API method, the skin friction is solely dependent on cohesion. Effective stress changes with depth, and the API method disregards the effective stress effects in soil.

NAVFAC DM 7.2

$$f = \alpha \cdot C_u \cdot A_p$$

S = skin friction; C_u = cohesion; A_p = perimeter surface area of the pile

Table 5.1 α vs cohesion

Pile Type	Soil Consistency	Cohesion Range (kN/m²)	α
Timber and concrete piles	Very soft	0–12	0–1.0
	Soft	12–24	1.0–0.96
	Medium stiff	24–48	0.96–0.75
	Stiff	48–96	0.75–0.48
	Very stiff	96–192	0.48–0.33
Steel piles	Very soft	0–12	0.0–1.0
	Soft	12–24	1.0–0.92
	Medium stiff	24–48	0.92–0.70
	Stiff	48–96	0.70–0.36
	Very stiff	96–192	0.36–0.19

Source: NAVFAC DM 7.2

As in the API method, effective stress effects are neglected in the DM 7.2 method.

Bored Piles

Fleming et al. (1985)

$$f = \alpha \cdot C_u$$

f = unit skin friction; C_u = cohesion
$\alpha = 0.7$ for clays with $C_u < 25\,kN/m^2$ *(522 psf)*
$ = 0.35$ for clays with $C_u > 70\,kN/m^2$ *(1,460 psf)*

(Note: α value for bored piles is chosen to be 0.7 times the value for driven piles.)

5.4.2   Equations Based on Vertical Effective Stress → $f_{ult} = \beta \times \sigma'$

f_{ult} = ultimate skin friction
   β = skin friction coefficient based on effective stress
   σ' = effective stress

This equation ignores cohesion effects.

Burland (1973)

$$f = \beta \cdot \sigma_v{}'$$

f = unit skin friction; σ_v' = effective stress

$\beta = (1 - \sin \phi') \tan \phi' \, (OCR)^{0.5}$

OCR = over consolidation ratio of clay

The Burland method does not consider the cohesion of soil. One could argue that OCR is indirectly related to cohesion.

5.4.3   Equation Based on Both Cohesion and Effective Stress

A new method proposed by Kolk and Van der Velde (1996) considered both cohesion and effective stress to compute the skin friction of piles in clay soils.

Kolk and Van der Velde Method (1996)

The Kolk and Van der Velde method considers both cohesion and effective stress.

$$f_{ult} = \alpha \times S_u$$

f_{ult} = ultimate skin friction

α = skin friction coefficient obtained using the correlations provided by Kolk and Van der Velde. α is a parameter based on both cohesion and effective stress.

S_u = undrained shear strength (cohesion)

In the Kolk and Van der Velde equation, α is based on the ratio of undrained shear strength and effective stress. A large database of pile skin friction results was analyzed and correlated to obtain α values.

Table 5.2 Skin friction factor

S_u/σ'	0.2	0.3	0.4	0.5	0.6	0.7	0.8	0.9	1.0	1.1	1.2	1.3	1.4
α		0.95	0.77	0.7	0.65	0.62	0.60	0.56	0.55	0.53	0.52	0.50	0.49	0.48
S_u/σ'	1.5	1.6	1.7	1.8	1.9	2.0	2.1	2.2	2.3	2.4	2.5	3.0	4.0
α		0.47	0.42	0.41	0.41	0.42	0.41	0.41	0.40	0.40	0.40	0.4	0.39	0.39

Source: Kolk and Van der Velde (1996)

According to Table 5.2, α decreases when S_u increases and α increases when σ' (effective stress) increases.

Design Example 5.1

Find the skin friction of the 1-ft-diameter pile shown using the Kolk and Van der Velde method.

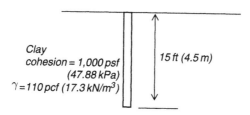

Clay
cohesion = 1,000 psf
(47.88 kPa)
γ = 110 pcf (17.3 kN/m³)

15 ft (4.5 m)

Figure 5.12   Pile in clay soil

Solution

The Kolk and Van der Velde method depends on both effective stress and cohesion. Effective stress varies with the depth. Hence, you should obtain the average effective stress along the pile length and in this case, obtain the effective stress at the midpoint of the pile.

Effective stress at midpoint of pile $= 110 \times 7.5$ psf $= 825$ psf   *(39.5 kPa)*
Cohesion $= S_u = 1,000$ psf   *(47.88 kPa)*
$S_u/\sigma' = 1,000/825 = 1.21$

From the table provided by Kolk and Van der Velde $\alpha = 0.5$ for $S_u/\sigma' = 1.2$.
Hence, use $\alpha = 0.5$.
Ultimate unit skin friction $= \alpha \times S_u = 0.5 \times 1,000 = 500$ psf   *(23.9 kPa)*
Ultimate skin friction of the pile $= 500 \times (\pi \times d \times L)$
$$= 500 \times \pi \times 1 \times 15 \text{ lbs}$$
$$= 23,562 \text{ lbs}   (104.8 \, kN)$$

References

American Petroleum Institute, "Recommended Practice for Planning, Designing and Constructing Fixed Offshore Platforms," API RP2A, 15th ed, 1984.
Burland, J.B., "Shaft Friction of Piles in Clay—A Simple Fundamental Approach," *Ground Eng.* 6, No. 3, 30–42, 1973.

Fleming, W.G.K., et al., *Piling Engineering*, Surrey University Press, New York, 1985.

Kolk, H.J., and Van der Velde, A., "A Reliable Method to Determine Friction Capacity of Piles Driven into Clays," Proc. Offshore Technological Conference, Vol. 2, Houston, TX.

McCammon, N.R, and Golder, H.G., "Some Loading Tests on Long Pipe Piles," *Geotechnique* 20-No 2 (1970).

NAVFAC DM 7.2, *Foundation and Earth Structures*, U.S. Department of the Navy, 1984.

Seed, H.B., and Reese, L.C., "The Action of Soft Clay Along Friction Piles," *Proc. Am. Soc. of Civil Engineers.*, 81, Paper 842.

5.5   Bored Piles in Clay Soils

5.5.1   Skin Friction in Clay Soils

As mentioned in previous chapters, attempts to correlate skin friction with undrained shear strength were not very successful. Better correlation was found with the skin friction and (Su/σ') ratio, "Su" being the undrained shear strength and σ' being the effective stress.

What will happen to a soil element when a hole is drilled?

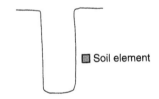

Figure 5.13   Soil element near a drilled hole

The following changes occur in a soil element near the wall after the drilling process:

- The soil element shown will be subjected to a stress relief. Undrained shear strength decreases due to the stress relief.

- Reduction of undrained shear strength reduces the skin friction as well.

5.5.2   Computation of Skin Friction in Bored Piles

- The same procedure used for driven piles can be used for bored piles, but with less undrained shear strength. The question is how much reduction should be applied to the undrained shear strength for bored piles?

- Undrained shear strength may decline as much as 50% due to the stress relief in bored piles. On the other hand, measured undrained shear strength is already reduced due to the stress relief that occurs when the sample is removed from the ground. By the time the soil sample reaches the laboratory, the soil sample has undergone stress relief and measured value already indicates the stress reduction. Considering these two aspects, reduction of 30% in the undrained shear strength is realistic for bored piles.

References

Kolk H.J., and Van der Velde A., *"A Reliable Method to Determine Friction Capacity of Piles Driven into Clays,"* Proc. Offshore Technological Conference, 1996.

O' Neill, M.W., "Side Resistance in Piles and Drilled Shafts," *ASCE Geotechnical and Geoenviroenmental Eng. J.,* Jan. 2001.

Meyerhoff, G.G., "Bearing Capacity and Settlement of Pile Foundations," *J. of Geotech. Eng, ASCE,* 102(3), 195–228, 1976.

Design Example 5.2 (Single Pile in a Uniform Clay Layer)

Find the capacity of the pile shown in Figure 5.14. The length of the pile is 10 m, and the diameter of the pile is 0.5 m. Cohesion of the soil is 50 kPa, and $\alpha = 0.75$. Note that the groundwater level is at 2 m below the surface.

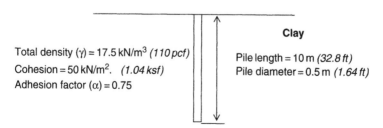

Figure 5.14   Single pile in a uniform clay layer

Solution

STEP 1: Find the end bearing capacity.

End bearing capacity in clay soils $= 9 \cdot c \cdot A$
$c = $ cohesion $= 50 \text{ kN/m}^2$
$Nc = 9$
$A = \text{л} \times D^2/4 = \text{л} \cdot 0.5^2/4 \text{ m}^2 = 0.196 \text{ m}^2$
Ultimate end bearing capacity $= 9 \times 50 \times 0.196 = 88.2 \text{ kN}$ *(19.8 kips)*

STEP 2: Find the skin friction.

Ultimate skin friction $= S_u = \alpha \times c \cdot A_p$
Ultimate skin friction $= 0.75 \times 50 \times A_p$
$A_p = $ perimeter surface area of the pile $= \text{л} \times D \times L = \text{л} \times 0.5 \times 10 \text{ m}^2$
$A_p = 15.7 \text{ m}^2$
Ultimate skin friction $= 0.75 \times 50 \times 15.7 = 588.8 \text{ kN}$ *(132 kips)*

STEP 3: Find the ultimate capacity of the pile.

Ultimate pile capacity $=$ ultimate end bearing capacity $+$ ultimate skin friction
Ultimate pile capacity $= 88.2 + 588.8 = 677 \text{ kN}$ *(152 kips)*
Allowable pile capacity $=$ ultimate pile capacity/FOS
Assume a factor of safety of 3.0.
Allowable pile capacity $= 677/3.0 = 225.7 \text{ kN}$ *(50.6 kips)*

Note: The skin friction was much higher than the end bearing in this situation.

Design Example 5.3 (Single Pile in a Uniform Clay Layer with Groundwater Present)

Find the allowable capacity of the pile shown. Pile diameter is given to be 1 m, and the cohesion of the clay layer is 35 kPa. Groundwater is 2 m below the surface. Find the allowable capacity of the pile.

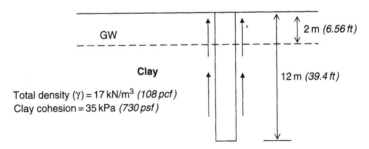

Figure 5.15 Skin friction in a pile

Solution

Unlike sands, groundwater does not affect the skin friction in clayey soils.

STEP 1: Find the end bearing capacity.

End bearing capacity in clay soils $= 9 \cdot c \cdot A$
  $c = $ cohesion $= 35\,\text{kN/m}^2$
  $N_c = 9$
  $A = \pi \times D^2/4 = \pi \cdot 1^2/4\,\text{m}^2 = 0.785\,\text{m}^2$
Ultimate end bearing capacity $= 9 \times 35 \times 0.785 = 247.3\,\text{kN}$ (*55.5 kips*)

STEP 2: Find the skin friction.

$$\text{Ultimate skin friction} = S_u = \alpha \times c \cdot A_p$$

Adhesion factor, α, is not given. Use the method given by the American
  Petroleum Institute (API).
$\alpha = 1.0$ for clays with cohesion $= <25\,\text{kN/m}^2$
$\alpha = 0.5$ for clays with cohesion $= >70\,\text{kN/m}^2$

Since the cohesion is $35\,\text{kN/m}^2$, interpolate to obtain the adhesion
factor (α).

$$25 \text{ ----------- } 1.0$$
$$35 \text{ ----------- } X$$
$$70 \text{ ----------- } 0.5$$

$$(X - 1.0)/(35 - 25) = (X - 0.5)/(35 - 70)$$
$$(X - 1.0)/10 = (X - 0.5)/-35$$
$$-35X + 35 = 10X - 5$$
$$45X = 40$$
$$X = 0.89$$

Hence, α at $35\,\text{kN/m}^2$ is 0.89.
Ultimate skin friction $= 0.89 \times 35 \times A_p$
$A_p = $ perimeter surface area of the pile $= \pi \times D \times L = \pi \times 1.0 \times 12\,\text{m}^2$
$A_p = 37.7\,\text{m}^2$
Ultimate skin friction $= 0.89 \times 35 \times 37.7\,\text{kN}$
Ultimate skin friction $= 1{,}174.4\,\text{kN}$ (*264 kips*)

STEP 3: Find the ultimate capacity of the pile.

Ultimate pile capacity = ultimate end bearing capacity + ultimate
skin friction

Ultimate pile capacity = 247.3 + 1,174.4 = 1,421.7 kN

Allowable pile capacity = ultimate pile capacity/FOS

Assume a factor of safety of 3.0.

Allowable pile capacity = 1,421.7/3.0 = 473.9 kN   (106.6 kips)

Design Example 5.4 (Computation of Skin Friction Using the Kolk and Van der Velde Method)

A 3-m sand layer is underlain by a clay layer with cohesion of
25 kN/m^2. Find the skin friction of the pile within the clay layer. Use
the Kolk and Van der Velde method. The density of both sand and clay
are 17 kN/m^3. Diameter of the pile is 0.5 m.

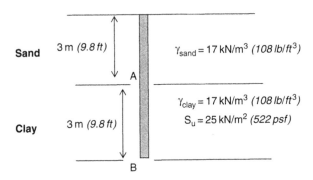

Figure 5.16   Pile in sand layer underlain by a clay layer

Solution

Find the skin friction at the top of the clay layer and bottom of the clay
layer. Obtain the average of the two values.

STEP 1:

At point A: $\sigma' = 3 \times 17 = 51$ kN/m^2 (1,065 lb/ft^2)
   $S_u = 25$ kN/m^2
$S_u/\sigma' = 25/51 = 0.50$

From the Kolk and Van der Velde table; $\alpha = 0.65$
At point B: $\sigma' = 6 \times 17 = 102$ kN/m^2
$S_u = 25$ kN/m^2
$S_u/\sigma' = 25/102 = 0.25$
From the Kolk and Van der Velde table; $\alpha = 0.86$

STEP 2

Ultimate skin friction at point A: $f_{ult} = \alpha \times S_u = 0.65 \times 25 = 16$ kN/m^2
Ultimate skin friction at point B: $f_{ult} = \alpha \times S_u = 0.86 \times 25 = 21$ kN/m^2
Assume the average of two points to obtain the total skin friction.
Average $= (16 + 21)/2 = 18.5$ kN/m^2
Total skin friction $= 18.5 \times$ (perimeter) $\times$ length $= 18.5 \times (\pi \times 0.5) \times 3$
       $= 87$ kN *(19.6 kips)*

Table 5.3   Summary of equations

	Sand	Clay
Pile end bearing capacity	$N_q \cdot \sigma_v \cdot A$	$9 \cdot c \cdot A$
Pile unit skin friction	$K \times \sigma_v \times \tan \delta \times A_p$	$\alpha \cdot c \cdot A_p$

$A =$ bottom cross-sectional area of the pile
$A_p =$ perimeter surface area of the pile
$\sigma_v =$ vertical effective stress
$N_q =$ terzaghi bearing capacity factor
$c =$ cohesion of the soil
$K =$ lateral earth pressure coefficient
$\alpha =$ adhesion factor
$\delta =$ soil and pile friction angle

5.6   Case Study: Foundation Design Options

D'Appolonia and Lamb (1971) describe construction of several buildings at MIT. The soil conditions of the site are given in Figure 5.17. Different foundation options were considered for the building.

5.6.1   General Soil Conditions

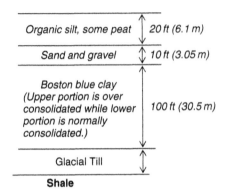

- Organic silt was compressible and cannot be used for shallow foundations.
- Sand and gravel were medium dense and can be used for shallow foundations.
- Upper portion of Boston blue clay was over consolidated. Lower portion was normally consolidated.
- Normally consolidated clays settle appreciably more than over consolidated clays.
- Glacial till can be used for end bearing piles.

Figure 5.17   Soil profile

Note: All clays start as normally consolidated clays. Over consolidated clays had been subjected to higher pressures in the past than existing in situ pressures. This happens mainly due to glazier movement, fill placement, and groundwater change. On the other hand, normally consolidated soils are presently experiencing the largest pressure ever. For this reason, normally consolidated soils tend to settle more than over consolidated clays.

Foundation Option 1: Shallow Footing Placed on Compacted Backfill

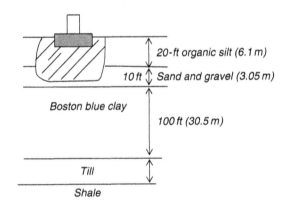

Figure 5.18   Shallow foundation option

Construction Procedure

- Organic silt was excavated to the sand and gravel layer.
- Compacted backfill was placed, and footing was constructed.
- This method was used for light loads.
- It has been reported that Boston blue clay would settle by more than 4 in. when subjected to a stress of 400 psf (Aldrich, 1952). Hence, engineers had to design the footing so that the stress on Boston blue clay was less than 400 psf.

Foundation Option 2: Timber Piles Ending on Sand and Gravel Layer

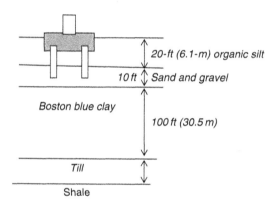

20-ft (6.1-m) organic silt

10 ft Sand and gravel

Boston blue clay

100 ft (30.5 m)

Till

Shale

Figure 5.19   Short timber pile option

Construction Procedure

- Foundations were placed on timber piles ending in a sand and gravel layer.
- Engineers had to make sure that the underlying clay layer was not stressed excessively due to piles.

This option was used for light loads.

Foundation Option 3: Timber Piles Ending in Boston Blue Clay Layer

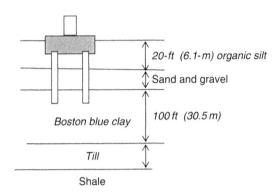

Figure 5.20   Long timber pile option

Construction Procedure

- Pile foundations were designed on timber piles ending in Boston blue clay.

- This method was found to be a mistake, since huge settlements occurred due to consolidation of the clay layer.

Foundation Option 4: Belled Piers Ending in Sand and Gravel

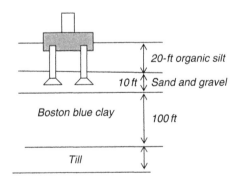

Figure 5.21   Belled pier option

Foundations were placed on belled piers ending in sand and gravel layer. It is not easy to construct belled piers in sandy soils. This option presented major construction difficulties.

Foundation Option 5: Deep Piles Ending in Till or Shale

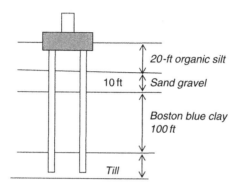

20-ft organic silt

10 ft   Sand gravel

Boston blue clay
100 ft

Till

Figure 5.22   Long piles to the till

- Foundations were placed on deep concrete pipe piles ending in till.

- These foundations were used for buildings with 20 to 30 stories.

- Their performance was found to be excellent. Settlement readings in all buildings were less than 1 in.

- Closed-end concrete filled pipe piles were used. These piles were selected over H-piles because of their lower cost.

- During pile driving, adjacent buildings underwent slight upheaval. After completion of driving, buildings started to settle. (Great settlement in adjacent buildings within 50 ft of piles was noticed.)

- Measured excess pore water pressures exceeded 40 ft of water column, 15 ft away from the pile. Excess pore pressures dropped significantly after 10 to 40 days.

- Due to upheaval of buildings, some piles were pre-augered down to 15 ft, prior to driving. Pre-augering reduced the generation of excess pore water pressures. In some cases, excess pore water pressures were still unacceptably high.

- Piles in a group were not driven at the same time. After one pile was driven, sufficient time was allowed for the pore water pressure to dissipate prior to driving the next pile.

- Another solution was to drive the pipe piles open end and then to clean out the piles. This option was found to be costly and time consuming.

Foundation Option 6: Floating Foundations Placed on Sand and Gravel (Rafts)

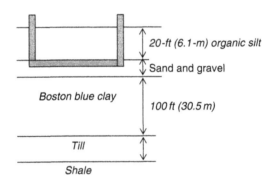

Figure 5.23   Floating (mat) foundation option

- Mat or floating foundations were placed on sand and gravel.

- Settlement of floating foundations was larger than deep piles (Option 5).

- Settlement of floating foundations varied from 1.0 to 1.5 in. Performance of rafts was inferior to deep piles. Average settlements in raft foundations were slightly higher than deep footings.

- There was a concern that excavations for rafts could create settlements in adjacent buildings. This was found to be a false alarm since adjacent buildings did not undergo any major settlement due to braced excavations constructed for raft foundations.

- When the cost of driving deep piles is equal to the cost of raft foundations, it is desirable to construct raft foundations, since rafts would give a basement.

- A basement may not be available in the piling option, unless it is specially constructed with additional funds.

Foundation Type	Bearing Stratum	No. of Buildings	Average No. of Stories	Settlement
Timber piles	Boston blue clay	27	1 to 6	1 to 16 in.
Belled caissons	Sand and gravel	22	1 to 8	1 to 3 in.
Raft foundations	Sand and gravel	7	6 to 20	1 to 1.5 in.
Deep piles	Shale or Till	5	6 to 30	0.5

References

Aldrich, H.P., *Importance of the Net Load to Settlement of Buildings in Boston*, Contributions of Soil Mechanics, Boston Soc. of Civil Engineers, 1952.

D'Appolonia, D.J., and Lamb, W.T., Performance of Four Foundations on End Bearing Piles, *ASCE, J. of Soil Mechanics and Foundation Eng.*, January, 1971.

Horn, H.M., and Lamb, T.W., "Settlement of Buildings on the MIT Campus," *ASCE, J. of Soil Mechanics and Foundation Eng.*, September, 1964.

5.7   Maximum Allowable Pile Loads

The following guidelines can be used when recommending pile capacities (*NYC Building Code*). These guidelines are provided for reference purposes only. Practicing engineers should refer to local building codes for applicable regulations.

- Open-end pipe piles greater than 18 in. in diameter, bearing on sound rock to intermediate rock. *Maximum allowable capacity 250 tons.*

- Open-end pipe piles smaller than 18 in. in diameter, bearing on sound rock to intermediate rock. *Maximum allowable capacity 200 tons.*

- Closed-end pipe piles, H-piles, and cast-in-place concrete piles bearing on sound rock to intermediate rock. *Maximum allowable capacity 150 tons.*

- Closed-end pipe piles, cast in place concrete piles and compacted concrete piles bearing on soft rock. *Maximum allowable capacity 60 tons.*

- Open-end pipe piles and H-piles bearing on soft rock. *Maximum allowable capacity 80 tons.*

- Timber piles bearing on any type of rock. *Maximum allowable capacity 25 tons.*

6

Pile Design: Special Situations

6.1 Timber Pile Design

6.1.1 Quality of Timber Piles

The engineer needs to make sure that the timber is of good sound quality, free from decay, and without damage during transportation and insect attack.

6.1.2 Knots

Timber piles contain knots and need to be observed. Local codes may provide guidelines on acceptable and nonacceptable knots. Acceptable knot size is dependent on the type of timber. Typically, knots greater than 4 ins are considered unacceptable. Knots grouped together (cluster knots) may be undesirable, and such piles should be rejected.

6.1.3 Holes

Timber piles may contain holes that are larger than acceptable size. Local codes should be referred for the acceptable size of holes. Typically, holes greater than ½ in. may be considered undesirable.

6.1.4 Preservatives

No portion of the pile should be exposed above the groundwater unless the pile is treated with a preservative. Untreated piles usually last a very long time below groundwater level. Piles are susceptible to decay above groundwater level.

Creosote is widely used, and other preservatives such as chlorophenols and napthelenates are also used to treat piles. Some preservatives are not suitable for piles located in marine environments. The pile designer should investigate the preservative type for the environment where the piles are located. Timber species also affects the preservative type.

Piles in Marine Environments

A study has been done by Grall (1992) of many bacteria, algae, and other unicellular organisms that get attached to piles in the ocean. Following is a summary of Grall's account.

When a pile is driven to the bottom of the bay, life takes up residence almost immediately. Bacteria, algae and protozoans cover the submerged surface. This "slime" provides a foothold for larger creatures to attach themselves in succession. In the summer ivory barnacles, bryzoans and sun sponges get attached themselves to the pile. Bright patches of algae such as sea lettuce soon arrive, followed by hydroids and bulbous sea squirts. Tube builder amphipods construct tunnels of mud and detritus for protection and for a niche on the crowded pile. Almost every underwater part of the piling is covered with various species each looking for food, shelter and a place to propagate.

Eventually the pile would fail and drop to the ocean floor taking the animals and the structure it supports as well. Hence, timber piles need to be preserved with extra care when used in marine environments.

Allowable Stresses in Timber

Allowable stress in timber piles is dependent on the timber species. ASTM D25 provides allowable stress in various timber species. A few examples are as follows.

Timber Species	Compression Parallel to Grains	Bending Stress	Modulus of Elasticity
Southern Pine	1,200 psi	2,400 psi	1.5×10^6 psi
Douglas Fir	1,250 psi	2,450 psi	1.5×10^6 psi

Figure 6.1   Compression parallel to grains

Straightness Criteria: Piles are made of timber and may not be as straight as a steel H-pile. Straightness criteria for timber piles should be established by the engineer. Local codes may provide the minimum criteria required. Rule-of-thumb methods such as drawing a straight line from one end to the other to make sure that the line is within the pile is popular.

Allowable Working Stress for Round Timber Piles

The allowable working stress of round timber piles depends on the wood species. The following table is provided by the American Association of State Highway and Transportation Officials (AASHTO).

Timber Species	Allowable Unit Working Stress Compression Parallel to Grain for Normal Duration of Loading *(psi)*	Timber Species	Allowable Unit Working Stress Compression Parallel to Grain for Normal Duration of Loading *(psi)*
Ash, White	1,200	Hickory	1,650
Beech	1,300	Larch	1,200
Birch	1,300	Hard Maple	1,300
Chestnut	900	Oak (red and white)	1,100
Southern Cypress	1,200	Pecan	1,650
Tidewater red Cypress	1,200	Pine, Lodgepole	800
Douglas Fir, inland	1,100	Pine, Norway	850
Douglas Fir coast type	1,200	Pine, Southern	1,200
Elm, rock	1,300	Pine, Southern dense	1,400
Elm, soft	850	Poplar, yellow	800
Gum (black and red)	850	Redwood	1,100
Eastern Hemlock	800	Spruce, Eastern	850
West coat Hemlock	1,000	Tupelo	850

Timber Pile Case Study

The site contained 5 ft of loose, fine sand followed by 15 ft of soft clay. Fine sand had an average SPT (N) value of 3, and the cohesion of soft clay was found to be 500 psf. Medium-dense coarse sand was encountered below the clay layer, which had an average SPT (N) value of 15. Groundwater was found to be at 8 ft below the surface. Due to loose,

fine sand and soft clay, shallow foundations were considered to be risky. A decision was made to drive piles to the medium-dense sand layer. Timber piles were selected owing to the lesser cost compared to other types of piles. Static analysis was conducted with 8-in.-diameter, 30-ft-long piles.
Assume the density of all soils to be 110 pcf *(17.3 kN/m³)*.
    The main reasons for selecting timber piles are as follows.

• No boulders or any other obstructions were found in the overburden soil.

• Bearing stratum (medium dense sand) was within the reach of typical length of timber piles.

Piles were driven with a 20,000-lbs. ft hammer till 30 blows per foot. Conduct a static analysis and a dynamic analysis of the pile capacity.

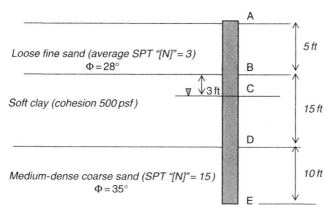

Figure 6.2   Soil profile and the pile

Static Analysis

STEP 1: Find the friction angle from SPT "N" values.

Loose, fine sand (N = 3), from table 1.1 $\Phi = 28°$
Coarse sand (N = 15), from table 1.1 $\Phi = 35°$

STEP 2: Find the end bearing capacity.

$$\text{Ultimate end bearing capacity } (Q_u) = q \times N_q \times A$$

q = Vertical effective stress at the bottom (tip) of the pile
N_q = Bearing capacity factor
A = Cross-sectional area

Effective stress at bottom of the pile $= 110 \times 5 + 110 \times 3 + (110 - 62.4$
$\times 12 + (110 - 62.4) \times 10$
$= 1,927 \text{ psf}$

Cross-sectional area of the pile $= \pi \times D^2/4 = \pi \times (8/12)^2/4 = 0.35 \text{ ft}^2$

$N_q = 50$ (Bottom of the pile is located in coarse sand with friction angle of 35°.
NAVFAC DM 7.2 provides $N_q = 50$.)

$$\text{Ultimate end bearing capacity} = q \times N_q \times A$$
$$= 1,927 \times 50 \times 0.35$$
$$= 33.7 \text{ kips}$$

STEP 3: Find the skin friction from A to B (loose, fine sand).

$$\text{Skin friction from A to B} = K \times q \times \tan \delta$$
$$\times \text{(perimeter surface area)}$$

q = Vertical effective stress at the midpoint of the section considered
K = Lateral earth pressure coefficient
$q = 110 \times 2.5 = 275 \text{ psf}$

$$K = (K_a + K_0 + K_p)/3$$

$K_a = \tan^2(45 - \Phi/2) = 0.36$
$K_0 = 1 - \sin\Phi = 1 - \sin 28 = 0.53$
$K_p = \tan^2(45 + \Phi/2) = 2.77$

$$K = (0.36 + 0.53 + 2.77)/3.0 = 1.22$$

δ for timber piles $= 3/4 \times \Phi$ (Ref. NAVFAC DM 7.2)

$$\delta = 3/4 \times \Phi = 3/4 \times 28 = 21°$$

Skin friction from A to B = K $\times$ q $\times$ tan δ $\times$ (perimeter surface area)
$$= 1.22 \times 275 \times \tan (21°) \times \pi \times (8/12) \times 5$$
$$= 1.35 \text{ kips } (6.0 \text{ } kN)$$

STEP 4: Find the skin friction from B to C (soft clay).

Skin friction from B to C (soft clay) = α $\times$ cohesion
$$\times \text{ (perimeter surface area)}$$

α = Adhesion factor = 0.96 (NAVFAC DM 7.2)
Cohesion = 500 psf

Skin friction from B to C (clay) = $0.96 \times 500 \times \pi \times 8/12 \times 3$
$$= 3.02 \text{ kips}$$

STEP 5: Find the skin friction from C to D (soft clay).

Skin friction from C to D (clay) = α $\times$ cohesion
$$\times \text{ (perimeter surface area)}$$

α = Adhesion factor = 0.96 (NAVFAC DM 7.2)
Cohesion = 500 psf

Skin friction from C to D (clay) = $0.96 \times 500 \times \pi \times 8/12 \times 12$
$$= 12.1 \text{ kips}$$

Note: One could have computed the skin friction from B to D directly in this situation since cohesion of saturated clay and unsaturated clay is the same.

STEP 6: Find the skin friction from D to E (coarse sand).

Skin friction from D to E = K $\times$ q $\times$ tan δ $\times$ (perimeter surface area)

q = Vertical effective stress at the midpoint of the section considered
K = Lateral earth pressure coefficient

$$q = 110 \times 8 + (110 - 62.4) \times 17 = 1,690 \text{ psf}$$
$$K = (K_a + K_0 + K_p)/3$$

$K_a = \tan^2(45 - \Phi/2) = 0.27$
$K_0 = 1 - \sin \Phi = 1 - \sin 35 = 0.43$
$K_p = \tan^2 (45 + \Phi/2) = 3.69$

$$K = (0.43 + 0.27 + 3.69)/3.0 = 1.46$$
$$\delta \text{ for timber piles} = 3/4 \times \Phi \text{ (Ref.NAVFAC DM 7.2)}$$
$$\delta = 3/4 \times 35 = 3/4 \times 35 = 26°$$

Skin friction from D to E $= K \times q \times \tan \delta \times$ (perimeter surface area)
$$= 1.46 \times 1,690 \times \tan(26°) \times \pi \times (8/12) \times 10$$
$$= 25.2 \text{ kips}$$

STEP 7:

End bearing capacity = 33.7 kips
Skin friction from A to B = 1.35
Skin friction from B to C = 3.02
Skin friction from C to D = 12.1
Skin friction from D to E = 25.2
Total ultimate pile capacity = 75.4 kips
Total allowable capacity = 25.1 kips (assuming a factor of safety of 3.0)

STEP 8: Hammer energy is given as 20,000 lbs. ft. The piles were driven till 30 blows per foot.
Engineering news record formula

R = Hammer energy/(s + 0.1)
R = Ultimate capacity of the pile
s = Distance per blow
30 blows per foot is equal to 1/30 ft per blow.
$s = 1/30 \text{ ft} = 0.033$
$R = 20,000/(0.033 + 0.1) \text{ lbs.} = 150 \text{ kips}$
Allowable capacity of the pile (dynamic analysis) = 150/3.0 = 50 kips
We obtained an allowable capacity of 25 kips from static analysis.

6.2 Case Study—Bridge Pile Design (Timber Piles)

6.2.1 Bridge Pile Design

Timber piles are rarely used for bridge abutments and piers today. This case study presents details of a timber pile project for bridge abutments and piers.

- Center pier is supported on 25 piles as shown above.

- 1-ft diameter timber piles, length = 50 ft pile cap = 15 × 15 ft; pile cap height = 3.5 ft

- Abutment is supported on 7 piles each with diameter of 1 ft. (See next page.)

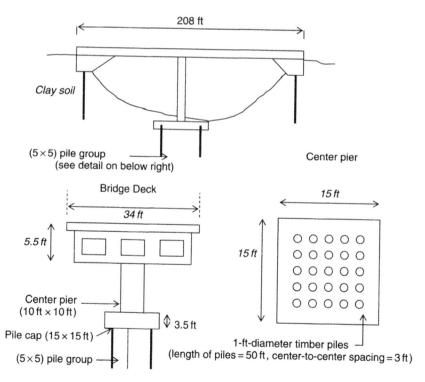

Figure 6.3   Bridge piles

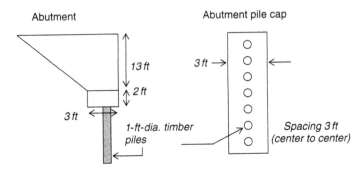

Figure 6.4   Bridge abutment

- *Soil Parameters*: Soil is mostly medium Stiff to stiff clay.
  Unconfined compressive strength = 2.3 tsf
  Average N value of the clay soil = 14 (Lower values occurred near the surface.)

- *Earthquake*: An earthquake of magnitude 6.4 caused significant shaking of the structure but no damage. (Peak horizontal and vertical accelerations measured during the earthquake were 0.5 g and 0.51 g, respectively.)

References

American Association of State Highway and Transportation Officials (AASHTO), Standard Specifications for Highway and Bridges, 1992.

Grall, G., "Pillar of Life," *Journal of the National Geographic Society* 114, 182, No. 1, July 1995.

Levine, M.B., and Scott, R.F., "Dynamic Response Verification of Simplified Bridge Foundation Model," *ASCE Geotechnical Eng. Journal* 15, No 2, February 1989.

6.3   Auger Cast Pile Design (Empirical Method)

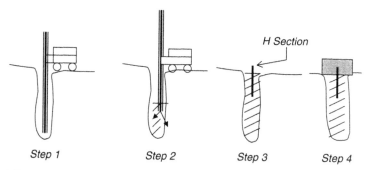

Figure 6.5   Auger cast pile construction. STEP 1: Auger a hole. STEP 2: Start pumping grout or concrete through the auger while lifting the auger. Grout should be pumped inside the grout. The auger should not be raised above the grout level. The auger should be lifted slowly so that the grout level is always above the tip of the auger. STEP 3: Concrete or grout the hole completely and place an "H" section or a steel rod to anchor the pile to the pile cap. STEP 4: Construct the pile cap.

6.3.1   Design Concepts

• The bearing capacity of auger cast piles is low compared to that similar of size driven piles.

• The volume of grout depends on the applied pressure. If grout is pumped at a higher pressure, more volume of grout will be pumped. In most cases, the volume of grout pumped is more than the volume of the hole.

• The grout factor is defined as the ratio of grout volume pumped to volume of the hole.

$$\text{Grout Factor} = \frac{\text{grout volume pumped}}{\text{volume of the hole}}$$

• Auger cast piles with high grout factors perform better than auger cast piles with low grout factors.

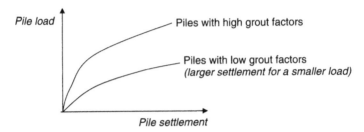

Figure 6.6   Pile settlement

- Piles with low grout factors have less skin friction since not much grout goes into the surrounding soil. On the other hand, when a higher grout pressure is applied, grout goes into the voids in the surrounding soil and develops a higher skin friction.

6.3.2 Auger Cast Pile Design in Sandy Soils

The ultimate strength of a pile depends on three factors:

- Skin friction

- End bearing

- Structural capacity of the pile

Computation of Skin Friction

The skin friction of auger cast piles is given by the following equation:

$fs = P_0' \times Ks \tan \delta$

fs = unit skin friction; P_0' = effective stress at the depth considered

Ks = lateral earth pressure coefficient;

δ = friction angle between grout and soil

Assume $P_0' \times Ks = \beta$. Hence, $fs = \beta \cdot \tan \delta$ tsf.

Table ACP 1   Shaft length vs. β

Shaft length (ft)	80	60	40	30	20	10
β		0.2	0.25	0.55	1.0	1.7	2.5

Source: Neely (1991).

- Intermediate values can be interpolated.

- Experiments indicate that the β for concrete and grout did not change significantly (*Montgomery 1980*).

- Research has not shown any variation of the β value based on the N value of sand. Hence, β factors are independent of N value of sand.

Computation of End Bearing:

After analyzing the vast number of empirical data, Neely (1991) provided the following equation:

$$q = 1.9 \, N < 75 \text{ tsf}$$

q = end bearing resistance (tsf); N = SPT value at the tip of the pile The "q" value should not be larger than 75 tsf.

References

Neely, W.J., "Bearing Capacity of Auger Cast Piles in Sand," *ASCE Geotechnical Eng. J.*, 1991.
Montgomery, M.W., "Prediction and Verification of Friction Pile Behavior," *Proc. ASCE Symp. on Deep Fdns,* Atlanta, GA, 274–287, 1980.

6.4   Capacity of Grouted Base Piles

Failure mechanisms of grouted base piles

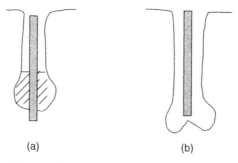

(a)   (b)

Figure 6.7   Grouted base pile failure. (a) Pile penetrates the grouted base and fails (punching failure). (b) Grouted base fails (splitting failure)

6.4.1   Structural Capacity of Grouted Base Piles

The structural capacity of grouted base piles is given by

$$Q = \sigma_c + 2\{\sigma_c/(\sigma_c - \sigma_t)\}^{1/2} \cdot P_a \quad (Kusakabe,\ 1994)$$

Q = pressure exerted by pile on the grouted base
σ_c = compressive strength of grout
σ_t = tensile strength of grout
P_a = ultimate inner pressure of hollow thick cylinder (The following equation for Pa is obtained from structural mechanics.)

$$P_a = \frac{\{1 - (a/b)^2\}\sigma_t + \{1 + (a/b)^2\}K_0 \cdot \gamma \cdot D}{2}$$

K_0 = earth pressure coefficient at rest = $(1 - \sin \Phi')$
   = Unit weight of soil

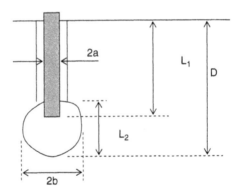

Figure 6.8   Grouted base pile diagram

Reference

Kusakabe, O., et al., "Structural Capacity of Precast Piles with Grouted Base," *ASCE Geotechnical Eng. J.*, August, 1994.

6.5   Case Study: Comparison between Bored Piles and Driven Piles

- A case study was conducted to compare the capacities of bored piles and driven piles in clay soils *(Meyerhof, G.G, 1953)*.

- Bored and driven piles were installed to a depth of 40 ft and load tested.

- Pile load test values were compared with theoretical computations.

    Clay Properties:

- The plastic limit and liquid limit of the clay were constant throughout the total depth.

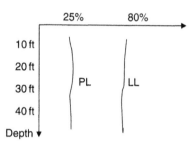

Figure 6.9   Atterberg limits

- Shelby tube samples were taken and tested for shear strength, under different conditions.

    1. Shelby tube samples were tested immediately after taking them (curve 1).

    2. Shelby tube samples were tested after remolding the clay (curve 2).

    3. Shelby tube samples were tested after 1 week (curve 3).

    4. Shelby tube samples were tested after softening them by adding water (curve 4).

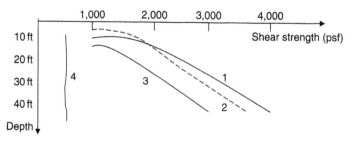

Figure 6.10   Shear strength vs. depth

- Samples tested immediately after taking them had the highest shear strength.
- Samples tested after adding water had the lowest shear strength.
- The shear strength of remolded samples was less than that of the ones tested immediately.
- Shear strength decreased when tested after one week. This was due to the creation of cracks and fissures during that time due to loss of moisture.

6.5.1   Results

- Driven piles had a higher capacity than bored piles.
- Load test values for bored piles were compatible with the shear strength values given by curve 4. The reason for this was assumed to be migration of water from wet concrete to surrounding clay. The increase of water content in clay decreased the shear strength value drastically.
- The solution to this problem is to use a dry concrete mix. In that case, compaction of concrete could be a problem.
- Pile load test values of driven piles were compatible with shear strength values given by curve 2. This was expected. When a pile is driven, the soil surrounding the pile is remolded.

Reference

Meyerhof, G.G., and Murdock, L.J., "Bearing Capacity of Bored Piles and Driven Piles in London Clay", *Geotechnique*, 3, 1953.

6.6   Case Study: Friction Piles

The capacity of a pile comes from skin friction and end bearing resistance. Piles that obtain their capacity mainly from skin friction are known as friction piles. Most engineers are more comfortable recommending end bearing piles than friction piles. This case study gives details of a project that included different types of friction piles.

6.6.1   Project Description (Blanchet, *et al.,* 1980)

The task was to design a bridge over Maskinonge River in Canada. Two abutments and four piers were deemed necessary for the bridge.

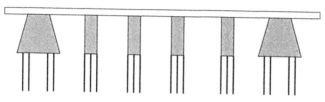

Figure 6.11   Friction piles in a bridge

Soil Condition at the Site

The top layers of soil were mostly sand, silt, and clayey silt. Below the top layers of soil, a very thick (175 ft) layer of silty clay was encountered. Below the silty clay layer, glacial till and shale bedrock were founded. It was clear that end bearing piles would be very costly for this site. If one were to design end bearing piles, they would be more than 220 ft in length.

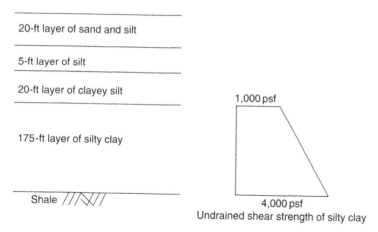

Figure 6.12   Soil profile

The trapezoid shows the distribution of undrained shear strength of the clay soil layer. Shear strength of the silty clay layer gradually increases from 1,000 psf to 4,000 psf.

The decision was made to design friction piles for this site.

Pile Types Considered

1. Tapered timber piles

2. Precast concrete piles (Herkules H-420, 2 segments)

3. Steel pipe piles (wall thickness 6.35 mm)

Load Test Data

Pile Type	Pile Diameter	Pile Length (ft)	Ultimate Load (tons)	Load Carried by 1 ft of Pile (tons/ft)
Tapered timber piles	Head = 14.5 in. Butt = 9 in.	50	78	1.56
Precast concrete piles	9 in.	78	65	0.83
Precast concrete piles	9 in.	120	95	0.80
Steel pipe piles	9 in	78	50	0.64

- Precast concrete piles with two different lengths were tested (78 ft and 120 ft).

- From load test data, it is clear that timber piles had the highest load-carrying capacity per foot of pile. This was mainly due to the taper of timber piles.

Load Settlement Curves

- Load settlement curves obtained during pile load tests were similar in nature, and one standard curve is shown in Figure 6.13.

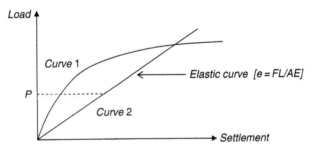

Figure 6.13   Load vs. settlement

- Above curve 2 is the elastic line shown for comparison purposes.

- The initial section of the pile load settlement curve is much steeper than the elastic curve.

- This means that for a given load, the pile would have a lesser settlement than the elastic compression. At a given load of P as shown, the settlement of the pile was almost half of the elastic settlement.

- This is due to skin friction between soil and pile. A portion of the load on the pile was absorbed by soil skin friction.

Settlement Values

- Settlement values obtained during pile load tests are given in the following table.

Pile Type	Timber Piles (50 ft long)	Steel Pipe Piles (78 ft long)	Precast Concrete Piles (78 ft long)	Precast Concrete Piles (120 ft long)
Settlement at 35 tons	3 mm	4 mm	2 mm	3 mm
Settlement at 45 tons	4 mm	6 mm	3 mm	4 mm
Settlement at 70 tons	8 mm	Fail	4 mm	8 mm

Settlement of piles occurs primarily for the following reasons.

- Elastic settlement of soil

- Elastic deformation of the pile

- Penetration of the pile into clay

- Long-term consolidation settlement of clay soil

At 35 tons, short precast concrete piles had the lowest settlement. Long precast concrete piles settled more than the short precast concrete piles with the same diameter. This is due to larger elastic compression in long piles.

In this situation, short piles may be more appropriate since, long piles would stress the soft silty clay underneath.

Pore Pressure Measurements

• Pore pressures were measured during pile driving. Pore pressure increase at a given depth was noticed to reach a peak value when the pile tip passed that depth.

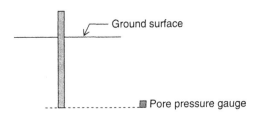

Figure 6.14 Pore pressure gauge location

• Pore pressure starts to decrease when the pile tip moves deeper.

Reference

Blanchet, C., et al. "Behavior of Friction Piles in Soft Sensitive Clays", *Canadian Geotechnical Eng. J.*, May 1980.

6.7 Open-End Pipe Pile Design—Semi-empirical Approach

Open-end pipe piles are driven in order to reduce driving stresses. During driving, a soil plug can develop inside the pipe pile.

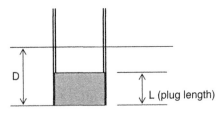

Figure 6.15 Plug length of can open end pile

Plug Ratio: Soil plug characteristics strongly affect the bearing capacity of open-end piles.

Plug Ratio $= L/D$

IFR (Incremental Filling Ratio): IFR is defined as the increase of the plug length with respect to increase of depth.

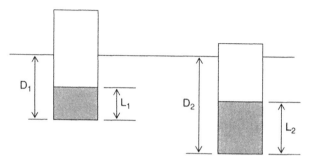

Figure 6.16 Development of plug

$$\text{IFR\%} = (L_2 - L_1)/(D_2 - D_1) \times 100$$

- If $L_1 = L_2$ then IFR $= 0$. In this case, soil plug length did not change due to further driving. This happens when the pile is driven through a soft soil strata.

- If $L_2 - L_1 = D_2 - D_1$, then IFR $= 1$. In this case, change in plug length is equal to change in depth. (When the pile is driven 1 ft, soil plug length increases by 1 ft.) This happens when the pile is driven through a hard soil stratum.

Measurement of IFR:

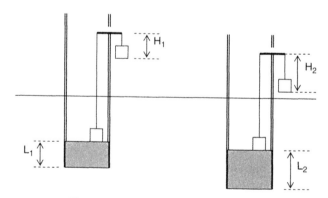

Figure 6.17 Measurement of IFR

- A hole is made through the pile as shown, and two weights are put in place.

- It is clear that $H_2 - H_1 = L_2 - L_1$. Hence, by measuring H_1 and H_2, it is possible to compute IFR.

  Correlation Between PLR and IFR:

  PLR = Plug length ratio = L/D
  IFR% = $(L_2 - L_1)/(D_2 - D_1) \times 100$
  Correlation $\rightarrow$ IFR% = $109 \times$ PLR $- 22$ *(Kyuho, 2003)*

6.7.1   End Bearing Capacity of Open-End Piles in Sandy Soils

The following equation was proposed by Kyuho (2003) to compute the end bearing capacity of open end pipe piles.

$$Q_b/(\alpha \cdot \sigma'_h) = 326 - 295 \times IFR\%/100$$

Q_b = ultimate end bearing capacity (same units as effective stress)
    = 1.0 for dense sands
    = 0.6 for medium sands
    = 0.25 for loose sands
σ'_h = horizontal effective stress = $K_0 \times \sigma'_v$
K_0 = earth pressure coefficient at rest = $(1 - \sin \Phi')$
σ'_v = vertical effective stress

6.7.2   Skin Friction of Open-End Pipe Piles in Sandy Soils

The following equation was proposed by Kyuho (2003) to compute the skin friction of open-end pipe piles.

$$f \text{ (unit skin friction)} = (7.2 - 4.8 \times PLR) \times (K_0 \cdot \sigma'_v \cdot \tan \delta) \cdot \beta$$

f = unit skin friction (Units same as σ'_v);
K_0 = lateral earth pressure coefficient = $(1 - \sin \phi')$
σ'_v = vertical effective stress
δ = friction angle between pile and soil
    = function of relative density
    = 1.0 for dense sands
    = 0.4 for medium sands
    = 0.22 for loose sands

6.7.3   Prediction of Plugging

It is important to predict the possibility of plugging during pile driving. Plugging of piles is dependent on denseness of soil. Sands with high relative density (Dr) have a higher tendency to plug than sands with low relative density.

Relative Density and Plugging

Relative Density (D_r) at Pile Tip	40	50	60	70	80	90	100
Internal Dia. of the Pile Required for Zero Plugging (m)	0.2	0.4	0.6	0.8	1.0	1.2	1.4

Source: Jardine and Chow (1996).

Example: If the relative density (D_r) of the soil at pile tip is 60%, find the minimum internal diameter required to avoid plugging.

Solution: 0.6 m (from the table above).

References

Jardine, R.J., and Chow, F.C., "New Design Methods for Offshore Piles", MTD Publication No: 96/103, Center for Petroleum and Marine Technology, London, 1996.

Kyuho, P., and Salgado, R., "Determination of Bearing Capacity of Open End Piles in Sand," *ASCE J. of Geotechnical and Geoenvironmental Eng.*, 2003.

Lethane, B.M., and Gavin, K.G., "Base Resistance of Jacked Pipe Piles in Sand," *J. of Geotechnical and Geoenvironmental Eng.*, June 2001.

Mayne, P., and Kulhawy, F.H., "K0–OCR Relationship in Soil," *ASCE J. of Geotechnical Eng.*, 1982.

O'Neill, M.W., and Raines, R.D., "Load Transfer for Pipe Piles in Highly Pressured Dense Sands," *J. of Geotechnical Eng.*, 1208-1226, 1991.

Randolph, et al., "One Dimensional Analysis of Soil Plugs in Pipe Piles," *Geotechnique*, 587–598, 1991.

6.8   Design of Pin Piles—Semi-empirical Approach

6.8.1   Theory

- A few decades ago, no engineer would have recommended any pile less than 9 in. in diameter. Today some piles can be as small as 4 in. in diameter.

- These small-diameter piles are known as *Pin Piles*. Other names such as *mini piles, micro piles, GEWI piles, Pali radice, root piles, and needle piles,* are also used to describe small-diameter piles.

Construction of a Pin Pile

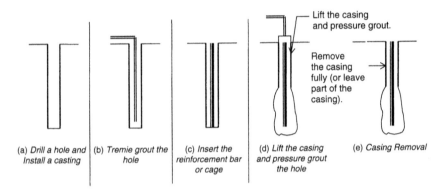

Lift the casing and pressure grout.

Remove the casing fully (or leave part of the casing).

(a) *Drill a hole and Install a casting* | (b) *Tremie grout the hole* | (c) *Insert the reinforcement bar or cage* | (d) *Lift the casing and pressure grout the hole* | (e) *Casing Removal*

Figure 6.18 Construction of pin piles.(a) Drill a hole using a roller bit or an auger. A casing is installed to stop soil from dropping into the hole. If the hole is steady, a casing may not be necessary. (b) Tremie grout the hole. (c) Insert the reinforcement bar or bars. In some cases, more than one bar may be necessary. (d) Lift the casing and pressure grout the hole. (e) Fully remove the casing. Some engineers prefer to leave part of the casing. This increases the strength and the cost.

Concepts to Consider

- **Fig. 6.18a:** Drilling the hole

  - Augering can be a bad idea for soft clays because augers tend to disturb the soil more than roller bits. This decreases the bond between soil and grout and lowers the skin friction. Drilling should be conducted with water. Drilling mud should not be used. Since casing is utilized to stop soil from falling in, drilling mud is not necessary. Drilling mud travels into the pores of the surrounding soil. When drilling mud occupies the pores, grout may not spread into the soil pores. This reduces the bond between grout and soil.

- **Fig. 6.18b:** Tremie Grouting

- After the hole is drilled, Tremie grout is placed. Tremie grout is a cement/water mix (typically with a water content ratio of 0.45 to 0.5 by weight).

- **Fig. 6.18c:** Placing Reinforcement Bars

  - A high-strength reinforcement bar (or number of bars wrapped together) or steel pipes can be used at the core. High-strength reinforcement bars specially designed for pin piles are available from Dywidag Systems International and William Anchors (leaders in the industry). These bars can be spliced easily, so that any length can be accommodated.

- **Fig. 6.18d:** Lifting the Casing and Pressure Grout

  - During the next step, the casing is lifted and pressure grouted. The pressure should be adequate to force the grout into the surrounding soil to provide a good soil/grout bond. At the same time, pressure should not be large enough to fracture the surrounding soil. Failure of surrounding soil would drastically reduce the bond between grout and soil. This would decrease the skin friction. Typically, grout pressure ranges from 0.5 MPa (10 ksf) to 1.0 MPa (1 MPa = 20.9 Ksf).

  - *Ground Heave:* In many instances, ground heave occurs during pressure grouting. This aspect needs to be considered during the design phase. If there are nearby buildings, action should be taken to avoid any grout flow into these buildings.

- **Fig. 6.18e:** Remove the Casing

  - During end of pressure grouting, the casing is completely removed. Some engineers prefer to leave part of the casing intact. Obviously, this increases the cost. The casing increases the rigidity and strength of the pin pile. Casing increases the *lateral resistance* of the pile significantly. Furthermore it is guaranteed that there is a pin pile with a diameter not less than the diameter of the casing. The diameter of a pin pile can be reduced due to soil encroachment into the hole. On some occasions it is possible for the grout to spread into the surrounding soil and create an irregular pile.

Figure 6.19 Irregular grout distribution

The grout can spread in an irregular manner as shown. This can happen when the surrounding soil is loose. The main disadvantage of leaving the casing in the hole is the additional cost.

Design of Pin Piles in Sandy Soils

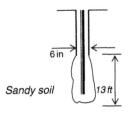

Figure 6.20 Pin pile diagram

Design Example:

Compute the design capacity of the pin pile shown (diameter 6 in). The surrounding soil has an average SPT (N) value of 15.

STEP 1: The ultimate unit skin friction in gravity grouted pin piles is given by

$$\tau = 21 \times (0.007N + 0.12) \text{ Ksf} \quad (\textit{Suzuki, et al., 1972}).$$

Note: The above equation has been developed based on empirical data.

N = SPT value; τ = unit skin friction in ksf
For N value of 15

$$\tau = 21 \times (0.007 \times 5 + 0.12) \text{ ksf}$$
$$\tau = 4.725 \text{ ksf}$$

The diameter of the pile = 6 in. and the length of the grouted section = 13 ft.

STEP 2:

Skin friction below the casing $= \pi \times (6/12) \times 13 \times 4.725$ kips
$$= 96.5 \text{ kips} = 48.2 \text{ tons.}$$

Note: Most engineers ignore the skin friction along the casing. The end bearing capacity of pin piles is not significant in most cases due to low cross sectional area. If the pin piles are seated on a hard soil or rock stratum, end bearing needs to be accounted.

Assume a factor of safety of 2.5.

Allowable capacity of the pin pile $= 48.2/(2.5) = 19$ tons

Note: (1) The above Suzuki's equation was developed for pin piles grouted using gravity. For pressure grouted pin piles, the ultimate skin friction can be increased by 20 to 40%. (2) (Littlejohn, 1970) proposed another equation.

$\tau = 0.21N$ ksf   (where N is the SPT value)
For $N = 15$, $\tau = 0.21 \times 15 = 3.15$ ksf
The above Suzuki's equation yielded $\tau = 4.725$ ksf.

For this case Littlejohn's equation provides a conservative value for skin friction. If Littlejohn's equation was used for the above pin pile:

Ultimate skin friction $= \pi \times (6/12) \times 13 \times 3.15 = 64.3$ kips
Allowable skin friction $= 64.3/2.5 = 25$ kips $= 12.5$ tons

Suzuki's equation yielded 19 tons for the above pin pile.

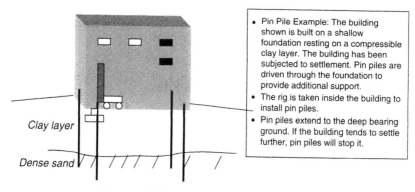

• Pin Pile Example: The building shown is built on a shallow foundation resting on a compressible clay layer. The building has been subjected to settlement. Pin piles are driven through the foundation to provide additional support.
• The rig is taken inside the building to install pin piles.
• Pin piles extend to the deep bearing ground. If the building tends to settle further, pin piles will stop it.

Clay layer

Dense sand

Figure 6.21   Pin piles in a building

References

Littlejohn, G.S., "Soil Anchors," Ground Engineering Conf., Institution of Civil Engineers, London, 33–44, 1970.

Suzuki et al., "Developments Nouveaux danles Foundations de Plyons pour Lignes de Transport THT du Japon," Conf. Int. de Grand Reseaux Electriques a haute tension, paper 21-01, 13 pp., 1972.

Xanthakos, P.P., Abramson, L.W, and Bruce, D.A., Ground Control and Improvement, John Wiley & Sons, New York, 1994.

6.9   Recommended Guidelines for Pile Design

The American Society of Civil Engineers (ASCE) has provided the following guidelines for pile foundation design. (ASCE 1997).

6.9.1   Steel Piles

- Steel pipe piles should have minimum yield strength of not less than 35,000 psi.

- Structural steel piles should conform to ASTM A36, ASTM A572, ASTM A588.

- Steel pipe piles should conform to ASTM A252.

- Steel encased cast-in-situ concrete piles should conform to ASTM A252, ASTM A283, ASTM A569, ASTM A570, or ASTM A611.

- The allowable design stress in steel should not be more than 35% of the minimum yield strength of steel.

Minimum Dimensions for Steel Pipe Piles

- Pipe piles should have a minimum outside diameter of 8 in.

- A minimum wall thickness of 0.25 in. is recommended for pipe diameters of 14 in. or less. Minimum wall thickness of 0.375 in. is recommended for pipe diameters greater than 14 in.

- Steel pipe piles with lesser wall thickness are allowed when the pipe piles are filled with concrete.

6.9.2   Concrete Piles

Reinforced Precast Concrete Piles

- Diameter or minimum dimension measured through the center should not be less than 8 in.
- Minimum 28-day concrete strength (f_c') = 4,000 psi.
- Minimum yield strength of re-bars = 40,000 psi.
- The allowable design stress in concrete should not be more than one-third of the minimum concrete strength.
- The allowable design stress in steel should not be more than 40% of the minimum yield strength of steel.

Prestressed Concrete Piles

- Diameter or minimum dimension measured through the center should not be less than 8 in.
- Minimum 28-day concrete strength = 4,000 psi.
- Minimum yield strength of re-bars = 40,000 psi.
- The effective prestress should not be less than 700 psi.
- The allowable axial design compressive stress applied to the full cross section should not exceed 33% of the specified minimum concrete strength minus 27% of the effective prestress force.

Concrete-Filled Shell Piles

- The diameter or minimum dimension measured through the center should not be less than 8 in.
- Minimum 28-day concrete strength = 3,000 psi.
- Minimum yield strength of re-bars = 40,000 psi.
- Thin shells less than 0.1 in. thick should not be considered as load-carrying members.
- The allowable design stress in concrete should not be more than one-third of the minimum concrete strength.
- The allowable design stress in steel should not be more than 40% of the minimum yield strength of steel.

Augered Pressure Grouted Concrete Piles

- Diameter should not be less than 8 in.
- Minimum 28-day concrete strength (f_c') = 3,000 psi.
- Minimum yield strength of rebars = 40,000 psi.
- The allowable design stress in concrete should not be more than one-third of the minimum concrete strength.
- The allowable design stress in steel should not be more than 40% of the minimum yield strength of steel.

6.9.3 Maximum Driving Stress

- Maximum driving stress for steel piles = $0.9\, f_y$ (for both tension and compression)

$$f_y = \text{yield strength}$$

- Maximum driving stress for timber piles = $2.5 \times (z)$

$$[z = \text{allowable design strength of timber piles}]$$

- Maximum driving stress for precast concrete piles = $0.85\, f_c'$ for compression

$$= 3(f_c')^{1/2} \text{ for tension}$$
$$f_c' = 28 \text{ day concrete strength.}$$

- Maximum driving stress for prestressed concrete piles = $(0.85\, f_c' - f_{pe})$ for compression

$$(f_{pe} = \text{Effective prestress force}).$$

- Maximum driving stress for prestressed concrete piles = $[3(f_c')^{1/2} + f_{pe}]$ for tension

$$(f_{pe} = \text{Effective prestress force}).$$

Reference

ASCE, *Standard Guidelines for the Design and Installation of Pile Foundations*, American Society of Civil Engineers, 1997.

6.10   ASTM Standards for Pile Design

6.10.1   Timber Piles

ASTM*

ASTM D25—Specification for round timber piles
ASTM D1760—Pressure treatment of timber products
ASTM D2899—Establishing design stresses for round timber piles

C 3 – Pile Preservative Treatment by Pressure Processes

C 18—Pressure treated material in marine environment
M 4—Standard for the care of preservative treated wood products

Steel H-Piles

ASTM A690—High-strength low-alloy steel H-piles and sheetpiles for use in marine environments.

Steel Pipe Piles

A 252—Specification for welded and seamless steel pipe piles

Pile Testing

ASTM D1143—Method of testing of piles under static axial compressive loads
ASTM D3689—Method of testing piles under static axial tensile loads
ASTM D3966—Testing piles under lateral loads
ASTM D4945—High strain dynamic testing of piles

---

*AWPA Standards for Timber Piles (American Wood Preservers Association, Woodstock, MD).

American Concrete Institute (ACI) Standards for General Concreting

ACI 304—Recommended practice for measuring, mixing, transportation, and placing concrete

ACI 305—Recommended practice for hot weather concreting

ACI 306—Recommended practice for cold weather concreting

ACI 308—Recommended practice for curing of concrete

ACI 309—Recommended practice for consolidation of concrete

ACI 315—Recommended practice for detailing of reinforced concrete structures

ACI 318—Requirements for structural concrete

ACI 347—Recommended practice for concrete formwork

ACI 403—Recommended practice for use of epoxy compounds with concrete

ACI 517—Recommended practice for atmospheric pressure steam curing of concrete

Design Stresses and Driving Stresses

Permissible design and driving stresses for various piles are given here. (Source: "Pile Driving Equipment," US Army Corps of Engineers, July 1997)

Pile Type	Authoritative Code	Design Stress (Permissible)	Driving Stress (Permissible)
Timber piles			
Douglas Fir		1.2 ksi	3.6 ksi
Red Oak		1.1 ksi	3.3 ksi
Eastern hemlock	ASTM D 25	0.8 ksi	2.4 ksi
Southern pine	ASTM D 25	1.2 ksi	3.6 ksi
Reinforced concrete piles (non-prestressed)	ACI 318 (concrete)	$0.33 f'_c$ (compression) f'_c = strength at 28 days. (Use gross cross-sectional area for compression and net concrete area for tension.)	$0.85 f'_c$ (compression) $3 [f'_c]^{1/2}$ (tension)
	ASTM A615 (reinforced steel)		

Pile Type	Authoritative Code	Design Stress (Permissible)	Driving Stress (Permissible)
Prestressed concrete	ACI 318 (concrete)	$0.33\, f'_c$ (compression) f'_c = strength at 28 days	For compression {$0.85\, f'_c$ – effective prestress}
Minimum effective prestress = 0.7 ksi Minimum 28-day concrete strength (f'_c) = 5.0 ksi	ASTM A615 (reinforced steel)	Use gross cross-sectional area for compression and net concrete area for tension.	For tension: {$3[f'_c]^{1/2}$ + effective prestress}
Steel pipe piles	ASTM A252 for steel pipe ACI 318 for concrete filling		$0.9\, f_y$ f_y = yield stress
Concrete cast-in-shell driven with mandrel	ACI 318 for concrete	$0.33\, f'_c$	Not applicable
Concrete cast-in-shell driven without mandrel	ASTM A36 for core ASTM A252 for pipe ACI 318 for concrete		$0.9\, f_y$ for steel shell (f_y = yield stress)
Steel HP section piles	ASTM A36		$0.9\, f_y$

6.11   Case Study: Prestressed Concrete Piles

In this case study, driving stresses generated inside a prestressed concrete pile are studied (*Mazen, 2001*).

Prestressed concrete pile

0.5-in.-dia. bars (6 bars total)
0.25-in.-dia. spiral reinforcement
355 mm
355 mm

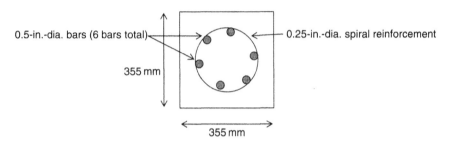

Figure 6.22 Prestressed concrete pile

Tensile strength of concrete is given by the following equation:

$$\text{Tensile strength} = 0.5(f_C')^{0.25}\ \text{MPa}$$

$$f_C' = \text{28-day concrete compressive strength (should be in MPa)}$$

$$E = \text{Young's modulus of concrete} = 4{,}700 \times (f'c)^{0.5}\ \text{MPa}$$

Soil Profile

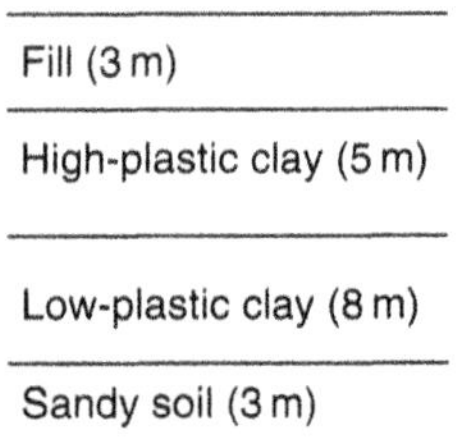

Fill (3 m)
High-plastic clay (5 m)
Low-plastic clay (8 m)
Sandy soil (3 m)

Figure 6.23 Soil profile

6.11.1 Pile Hammer Used: DELMAG D30-23 Diesel Hammer

- Pile will be subjected to tensile and compressive stresses during driving. The GRLWEAP program was used to calculate driving stresses.
- Computed maximum tensile stress occurring during driving when the stroke of the hammer was 1.5 m = 7 MPa.
- Maximum tensile stress occurred during driving when the stroke of the hammer was 2.1 m = 8.5 MPa.

- Tensile strength of concrete = 3.1 MPa

- The pile already had a prestress of 5.2 MPa (Prestress is a compressive stress.)

- Hence total tensile strength of the pile = (5.2 + 3.1) = 8.3 MPa < 8.5 MPa.

- Hence, a hammer stroke of 2.1 m cannot be used; use a hammer stroke of 1.5 m instead.

- Computed maximum compressive stress during driving = 28 MPa (from GRLWEAP program)

- 28-day compressive strength of concrete = 41 MPa

- The concrete was already prestressed to 5.2 MPa.

- Available compressive strength in concrete = (41 – 5.2) MPa = 35.8 MPa

- Compressive strength is greater than compressive stress occurring during driving (35.8 > 28).

- A hammer with a stroke of 1.5 m was used for this project.

Reference

Mazen, E.A., "Load tests of Pre-Stressed Pre-Cast Concrete and Timber Piles," *ASCE Geotechnical Eng. J.* 127, No. 12, December 2001.

6.12 Driving Stresses

- When a pile is driven, it is subjected to both tensile and compressive stresses. The pile needs to be designed to withstand driving stresses (compressive and tensile).

- Driving stresses are calculated using the wave equation.

- Computer programs such as GRLWEAP by Goble, Rausch, Likins, and Associates can be used to compute the driving stresses.

- Maximum tensile stress and maximum compressive stress during pile driving should be less than the tensile strength and compressive strength of the pile, respectively.

- Driving stresses are dependent on the stroke of the hammer and the driving resistance of soil.

Hammer energy = driving resistance × set
Hammer energy = weight of hammer × stroke

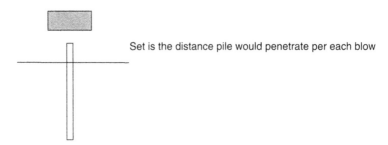

Set is the distance pile would penetrate per each blow

Figure 6.24   Pile driving

Example

For a 12-in. concrete pile, the following maximum tensile stresses are provided by the wave equation for a single-acting air hammer with a rating of 26,000 lb/ft.

Maximum Tensile Stress (lb/sq. ft)

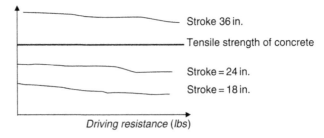

Stroke 36 in.

Tensile strength of concrete

Stroke = 24 in.
Stroke = 18 in.

Driving resistance (lbs)

Figure 6.25   Maximum tensile stress vs. driving resistance

According to the above data, the pile-driving operator should not use a stroke of 36 in.

6.13   Maximum Allowable Driving Stresses

AASHTO recommends following maximum allowable driving stresses.

- Steel piles-0.90 f_y (compression and tension)

  f_y = *steel yield stress*

- Concrete piles – 0.85 f_c' (compression)

  −0.70 f_y (for steel reinforcements under tension during driving)
  f_c' = 28-day concrete strength

- Prestressed concrete piles (normal environment)

  $[0.85 f_c' - f_{pe}]$ (compression)
  $(f_c' + f_{pe})^{1/2}$ (tension)
  f_{pe} = prestress force

- Prestressed concrete piles (corrosive environment)−f_{pe} (tension)

- Timber piles – 3 × allowable working stress (compression)

  3 × allowable working stress (tension)

Note: See earlier section, "Timber Piles," to obtain the allowable working stress for a given species of timber.

Reference

American Association of State Highway and Transportation Officials (AASHTO), *Standard Specifications for Highway Bridges*, 1992.

6.14   Uplift Forces

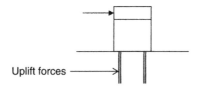

Figure 6.26   Uplift forces in-piles

- Outside piles in a building are usually subject to uplift forces.
- Total uplift capacity of a pile depends on the skin friction.

Is the skin friction in an uplift pile the same as the skin friction of a loaded pile?

- According to research data provided by Dennis and Olson (1983), there is no significant difference in skin friction between two situations.
- Contrary to Dennis and Olson's contention (1983), many engineers and academics believe that there is less skin friction during uplift than when the pile is loaded downward.

6.14.1   Uplift Due to High Groundwater

- Archimedes' theorem suggests that uplift force due to water is equivalent to the weight of water displaced.
- As shown in the figure, groundwater level has risen unexpectedly.
- This would increase the uplift force on the buiding structure. If the weight of the building is less than the uplift force due to buoyancy, then piles will be called in to resist the uplift force due to buoyancy.
- If the weight of the building is less than the buoyant forces acting upward, then the skin friction on piles will be facing down. The piles are acting as uplift piles.

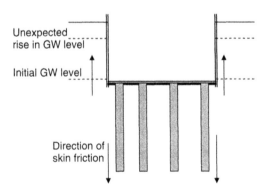

Figure 6.27   Uplift forces in a pile group

6.14.2   Uplift Forces Due to Wind

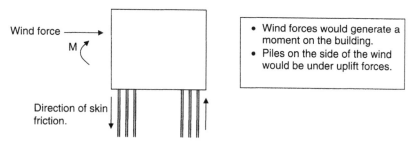

Figure 6.28   Uplift forces due to wind

- AASHTO recommends only one-third of the frictional resistance obtained using static equations for uplift piles.

- Uplift load test procedure can be found in ASTM D-3689.

Reference

American Association of State Highway and Transportation (AASHTO), *Standard Specifications for Highway Bridges*, 1992.

6.15   Load Distribution—Skin Friction and End Bearing

Generally, it is assumed that both skin friction and end bearing resistance occur simultaneously: in reality, this is not the case. When the pile is loaded, the total load is taken by skin friction at the initial stages. At small loads, end bearing resistance is almost negligible. When the load is increased, skin friction starts to increase without much increase in the end bearing. Load distribution between skin friction and end bearing is explained with an example below.

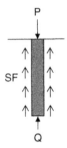

Figure 6.29   Skin friction and end bearing

- Assume the pile shown has an ultimate skin friction capacity (SF) of 600 kN and ultimate end bearing capacity (Q) of 400 kN.

- Hence total ultimate capacity of the pile is 1,000 kN. Assuming a factor of safety of 3.0, design capacity of the pile is 1,000/3.0 kN = 333 kN (74.8 kips)

- Load vs. settlement was recorded for the pile and shown in Figure 6.30. According to the figure, when the total load was 400 kN, settlement (s) was approximately 2 mm. Pile resistance due to skin friction was 390 kN, and end bearing was only 10 kN.

Total = 400 kN; skin friction = 390 kN; end bearing = 10 kN; s = 2 mm

End Bearing vs. Skin Friction (Typical Example)

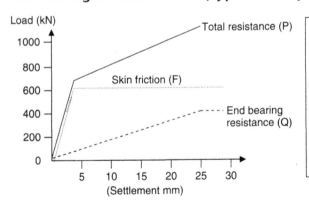

Figure 6.30   Load vs. settlement

- When P was increased to 615 kN, values were:

Total = 615 kN; skin friction = 600 kN; end bearing = 15 kN; s = 3 mm

When total load was increased beyond 615 kN, skin friction no longer increased. For an example, at a total load of 900 kN, the following values were obtained.

Total $= 900\,\text{kN}$; skin friction $= 600\,\text{kN}$; end bearing $= 300\,\text{kN}$; $s = 15\,\text{mm}$

- The pile had reached its ultimate skin friction at 600 kN. Skin friction will no longer increase. For this pile, the *ultimate skin friction* is developed at a settlement of 3 mm. Any additional load after the first 600 kN, would be taken fully by end bearing.

- Generally, piles generate skin friction at a very slight settlement, while full end bearing resistance occurs at a very high settlement.

Many engineers estimate the total ultimate capacity of the pile and divide it by a factor of safety. In some instances, little attention is paid to the development process of skin friction and end bearing. In the example given above, ultimate total resistance was 1,000 kN. Assuming a factor of safety of 3, one would obtain an allowable pile capacity of 333 kN.

At 333 kN, almost all the load is taken by skin friction (see the figure). At this load, end bearing resistance is negligible. Hence, one could question the logic of dividing the total ultimate resistance by a factor of safety.

7

Design of Caissons

7.1 Design of Caissons

The term caisson is normally used to identify large bored concrete piles. Caissons are constructed by drilling a hole in the ground and filling it with concrete. A reinforcement cage is placed prior to concreting. The diameter of caissons can be as high as 15 ft.

Other commonly used names to identify caissons are

a. Drilled shafts

b. Drilled caissons

c. Bored concrete piers

d. Drilled piers

A bell can be constructed at the bottom to increase the capacity. Caissons with a bell at the bottom are known as belled piers, belled caissons, or underreamed piers.

Brief History of Caissons

Many engineers recognized that large-diameter piles were required to transfer heavy loads to deep bearing strata. The obvious problem was the inability of piling rigs to drive these large-diameter piles.

The concept of excavating a hole and filling with concrete was the next progression. The end bearing of the caisson was improved by constructing a bell at the bottom.

Early excavations were constructed using steel rings and timber lagging. This idea was borrowed from the tunneling industry.

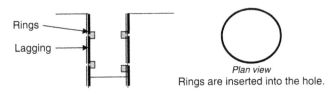

Figure 7.1   Early caissons

Machine Digging

With time, machine digging replaced hand digging. Two types of machines are in use.

• Auger types
• Bucket types

Augering is the most popular method used for excavating caissons.

7.2   Construction Methodology of Caissons (Dry Method)

(1) *A hole is drilled in the ground.*   (2) *A reinforcement cage is placed.*   (3) *The hole is concreted.*

Figure 7.2   Construction of caissons – dry method

Use of Casing

Both bucket and auger methods create problems when groundwater is present. Sidewall failure and difficulty of keeping the hole open are very common problems. The casing method is well suited for such situations, since the steel casing would prevent the sidewalls from falling.

The casing is inserted into the ground by pushing and rotating. Usually, the casing is removed during concreting. In some cases, the casing is left in the hole. Needless to say, this is an expensive proposition.

Use of Slurry

Slurry can also be used during drilling to keep the sidewalls from falling. Basically, two types of slurry are used.

- Polymer-based fluids
- Bentonite-based fluids

Polymer drilling fluids tend to increase the viscosity of the grout. Some polymers can lubricate the bored pile walls, causing less skin friction.

Belled Caissons

Belleld caissons are used to increase the bearing capacity of caissons. Because of the large area at the base, the end bearing capacity will increase. Belled caissons can be used to resist tension as well. The heavy weight and resistance at the bell provide additional capacity. Belling buckets are used to create the bell at the bottom.

Figure 7.3   Belled caisson

Integrity of Caissons

The following items are important for the integrity of a caisson.

- Control of concrete slump
- Avoiding sidewall failure

- Bottom clearing prior to concreting (In many occasions, soil from the sidewalls may fall to the bottom of the hole. This debris has to be cleaned prior to concreting.)
- Water in the bottom of the hole (If there is water in the bottom of the hole, it has to be removed prior to concreting. If there is a significant amount of water, then the Tremie method should be used for concreting).

Examining Caissons for Defects

If a caisson is suspected of being defective, the best way to inspect it is to construct a side test pit. This method may not be suitable for deep caissons. In such situations, the seismic wave method is used. A seismic wave is sent through the caisson, and the time lag for the return wave is measured. The return wave provides information regarding the integrity of the caisson.

Repairing Defective Caissons

If a caisson is found to be defective, the most common repair method is to drill holes and insert steel bars and grout. Another method is to provide an additional caisson to relieve the load on the defective caisson.

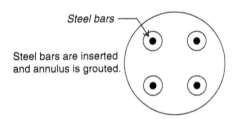

Figure 7.4   Steel bars in a caisson

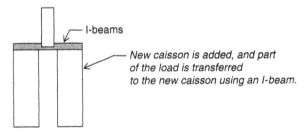

Figure 7.5   Adding a new caisson using an I-beam

References

Brown, D., "Effect of Construction on Axial Capacity of Drilled Foundation in Piedmont Soils," *ASCE J. of Geotechnical and Geoenvironmental Eng.* 128, No. 12, 2002.

Baker, C.N, and Kahn, F., "Caisson Construction," *ASCE J. of Soil Mechanics and Foundation Engineering,* 1971.

7.3   Caisson Inspection in Soil

The geotechnical engineer is responsible for inspection during construction of caissons. An experienced geotechnical engineer should carry out this inspection.

Items that need to be inspected are as follows:

- **Drilling Operation:** Soil can cave in during the drilling operation. The engineer should inspect the sides of the shaft and record the depth and extent of the caving of soil. The inspector should observe the movement of the auger. *The auger should go into the hole and come out of the hole without any disturbance.* Soil caving is extremely critical when there are nearby buildings. Soil movement could cause settlement in nearby buildings.

- **Water inflow:** The field engineer should inspect the walls of the shaft for water inflow. A powerful flashlight (or a mirror to direct sunlight) can be used to inspect the shaft wall. Water inflow could cause failure of sidewalls. Groundwater could mix with concrete and reduce the strength.

- **Shaft Centering:** The shaft should be centered as designed. The field engineer should drive four pegs close to the shaft prior to drilling. Then he should measure the distance to the shaft from these pegs to check the plumbness of the shaft.

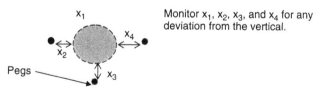

Figure 7.6   Construction monitoring of caissons

- **Bearing Stratum:** The field engineer should confirm the bearing stratum. The soil condition should be identified and compared to the soil conditions in nearby borings. If soil conditions are different, this information should be provided to the design engineer. Most projects require drilling to a certain soil stratum, and the field engineer is responsible for making sure that the required soil stratum is reached. If not, he or she should direct the contractor to dig deeper. In some cases, the field engineer may have to enter the hole. The field engineer should make sure that OSHA safety requirements are met prior to entering the hole. If the hole is drilled without a casing, a monitoring casing (or protection casing) should be utilized for the inspection purposes.

- **Construction of the Bell:** Under-reamed shafts (shafts with a bell at the bottom) need special attention. The field engineer should inspect the belling bucket to check the maximum angle it could excavate. Sometimes it may be necessary to go to the bottom of the hole to inspect the bell. In that case OSHA safety guidelines should be used prior to entering the hole. The field engineer should identify the soil stratum. Any presence of weak soil, water seepage, and angle and height of the bell should be recorded and provided to the design engineer.

- **Soil Caving:** If soil caving was noticed, the engineer has the option of recommending drilling mud. For loose sands and soft clays, a casing should be used to counter soil caving.

Preconstruction Meeting

A preconstruction meeting should be held among client, consultant, and the contractor prior to start of work. During this meeting the following items need to be discussed.

- **Unexpected Soil Conditions:** A payment procedure needs to be established if the caisson has to be drilled deeper due to unexpected weak soil conditions. Change orders for work delays due to boulders and other obstructions also need to be addressed. In most cases these items are included in contract documents. Finer details need to be clarified and agreed upon by all parties.

- **Field Inspector's Role:** The field inspector's role and his or her extent of authority need to be clarified. Theoretically, all changes need to be approved by the design engineer. This may not always be feasible. In such situations, the field engineer needs to be able to make decisions in oder to make changes to the size of the bell, depth of the shaft, usage of casings, concrete mixture, and construction procedures.

References

Baker, C.N., et al. "Drilled Shaft Inspector's Manual," *ASCE: The Association of Engineering Firms Practicing in the Geoscience,* 1989.

Reese, et al. "Behavior of Drilled Piers Under Axial Loading," *JGED, ASCE* 102, No. 5, May 1976.

American Society of Civil Engineers, "Design and Construction of Sanitary and Storm Sewers," *Manual of Practice* 37, New York, 1970 (5th Printing, 1982).

Bowles, J.E., *Foundation Analysis and Design,* 4th ed., McGraw-Hill, New York, 1988.

Dodson, R.D., *Storm Water Pollution Control,* McGraw Hill, New York, 1998.

Koppula, S.D., "Discussion: Consolidation Parameters Derived from Index Tests" *Geotechnique,* 36, No. 2, June 1986.

Nagaraj, T.S., and Murthy, B.R., "A Critical Reappraisal of Compression Index," *Geotechnique,* 36, No. 1, March 1986.

Reese, L., and Tucker, K.L., "Bentonite Slurry Constructing Drilled Piers,"*ASCE Drilled Piers and Caissons,* 1985.

Terzaghi, K, and Peck, R.B., *Soil Mechanics in Engineering Practice,* John Wiley and Sons, New York, 1967.

The equations for caissons in clay soils are similar to those for piles.

Different Methods

Reese et al. (1976): ultimate caisson capacity = ultimate end bearing capacity + ultimate skin friction

$$P_u = Q_u + S_u$$

P_u = ultimate capacity of the caisson
Q_u = ultimate end bearing capacity
S_u = ultimate skin friction

$$\text{End bearing capacity } (Q_u) = 9 \cdot c \cdot A$$

c = cohesion;
A = cross-sectional area

$$\text{Skin friction} = \alpha \cdot c \cdot A_p$$

"α" is obtained from Table 7.1.
  c = cohesion
A_p = perimeter surface area

Table 7.1   "α" value

Caisson Construction Method	α (for soil to concrete)	Limiting Unit Skin Friction (f)
		kPa
Uncased Caissons		
(1) Dry or using lightweight drilling slurry	0.5	90
(2) Using drilling mud where filter cake removal is uncertain (see note 1)	0.3	40
(3) Belled piers on about same soil as on shaft sides		
(3.1) By dry method	0.3	40
(3.2) Using drilling mud where filter cake removal is uncertain	0.15	25
(4) Straight or belled piers resting on much firmer soil than around shaft	0.0	0
(5) Cased Caissons (see note 2)	0.1 to 0.25	

Source: Reese et al., (1976)

Note 1:
Filter Cake: When drilling mud is used, the layer of mud gets attached to the soil. This layer of mud is known as the filter cake. In some soils, this filter cake gets washed away during concreting. In some situations, it is not clear whether the filter cake will be removed during concreting. If the filter cake does not get removed during concreting, the bond between concrete and soil will be inferior. Hence, low α value is expected.

Note 2:
Cased Caissons: In the case of cased caissons, the skin friction would be between soil and steel casing. α is significantly low for steel—soil bond compared to concrete—soil bond.

Factor of Safety: Due to uncertainties of soil parameters such as cohesion and friction angle, the factor of safety should be included. Suggested factor of safety values in the literature range from 1.5 to 3.0. The factor of safety of a caisson affects the economy of a caisson significantly. Assume that the ultimate capacity was found to be 3,000 kN and that a factor of safety of 3.0 is used. In this case the allowable caisson capacity would be 1,000 kN. On the other hand, a factor of safety of 2.0 would give an allowable caisson capacity of 1,500 kN, or a 50% increase in the design capacity. A lower factor of safety value would result in a much more economical design. Yet, failure of a caisson can never be tolerated. In the case of a pile group, failure of one pile will probably not have a significant effect on the group. Caissons stand alone and typically are not being used as a group.

When selecting the factor of safety, the following procedure is suggested.

- Make sure enough borings are conducted so that soil conditions are well known in the vicinity of the caisson.

- Conduct sufficient unconfined strength tests on all different clay layers to obtain an accurate value of for cohesion.

- Make sufficient borings with SPT values to obtain the friction angle of sandy soils.

If the subsurface investigation program is extensive, a lower factor of safety of 2.0 could be justified. If not, a factor of safety of 3.0 may be the safe bet. As a middle ground, a factor of safety of 3.0 for skin friction and a factor of safety of 2.0 for end bearing can be used.

Weight of the Caisson

Unlike in piles, the weight of the caisson is significant and needs to be considered for the design capacity.

$$P_{allowable} = Q_u/FOS + S_u/FOS - W + Ws$$

$P_{allowable}$ = allowable column load
$\quad Q_u$ = ultimate end bearing capacity
$\quad S_u$ = ultimate skin friction
$\quad W$ = weight of the caisson
$\quad W_s$ = weight of soil removed

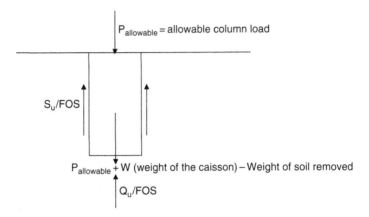

Figure 7.7   Forces acting on a caisson

Logically, the weight of the soil removed during construction of the caisson should be reduced from the weight of the caisson.

Ignore skin friction on top and bottom of the shaft

Reese and O'Neill (1988) suggest that the skin friction at the top 5 ft (1.5 m) of the caisson in clay be ignored since load tests show that the skin friction in this region is negligible. For the skin friction to mobilize, slight relative movement is needed between caisson and surrounding soil. The top 5 ft of the caisson do not have enough relative movement between the caisson and the surrounding soil. It has been reported that clay at the top 5 ft could desiccate and crack and result in negligible skin friction. At the same time, the bottom of the shaft will undergo very little relative displacement in relation to the soil as well. Hence, Reese and O'Neill suggest ignoring the skin friction at the bottom distance equal to the diameter of the caisson as well.

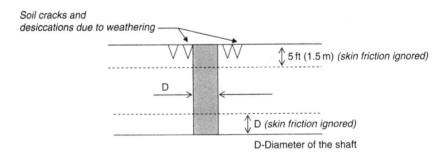

Figure 7.8   Skin friction acting on a caisson

Ignore the skin friction in

- Top 5 ft (1.5 m)
- Bottom distance equal to diameter of the shaft (D)

AASHTO Method

The American Association of State Highway and Transportation Officials (AASHTO) proposes the following method for caisson design in clay soils.

$$P_u = Q_u + S_u$$

P_u = ultimate capacity of the caisson
Q_u = ultimate end bearing capacity
S_u = ultimate skin friction

$$\text{End bearing capacity } (Q_u) = 9 \cdot c \cdot A$$

c = cohesion
A = cross-sectional area

$$\text{Skin friction} = \alpha \cdot c \cdot A_p$$

"α" adhesion factor is obtained from Table 7.2.
  c = cohesion
A_p = perimeter surface area

Table 7.2   AASHTO Skin friction coefficient (α)

Location along Drilled Shaft	Value of α	Maximum Allowable Unit Skin Friction
From ground surface to depth along drilled shaft of 5 ft	0	0
Bottom 1 diameter of the drilled shaft	0	0
All other portions along the sides of the drilled shaft	0.55	5.5 ksf (253 kPa)

It should be mentioned here that the AASHTO method has only one value for "α".

Design Example 7.1 (Caisson Design in Single Clay Layer)

Find the allowable capacity of a 2-m diameter caisson placed at 10 m below the surface. The soil was found to be clay with a cohesion of 50 kPa. The caisson was constructed using the dry method. The density of soil = 17 kN/m³. Groundwater is at 2 m below the surface.

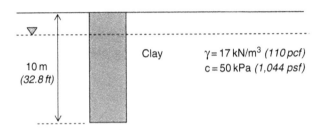

Figure 7.9   Caisson in a single clay layer

Solution

STEP 1: Find the end bearing capacity.

$$P_u = Q_u + S_u$$

End bearing capacity $(Q_u) = 9 \cdot c \cdot A$
End bearing capacity $(Q_u) = 9 \times 50 \times (\pi \cdot d^2)/4$

$$= 9 \times 50 \times (\pi \cdot 2^2)/4 = 1{,}413.8 \text{ kN} \quad (318 \text{ kips})$$

STEP 2: Find the skin friction.

The cohesion of soil and the adhesion factors may be different above groundwater level in the unsaturated zone. Unless the cohesion of the soil is significantly different above groundwater, such differences are ignored.

$$\text{Skin friction} = \alpha \cdot c \cdot A_p$$

$\alpha = 0.5$ from Table 7.1. The caisson was constructed using the dry method.)
$A_p = \pi \times d \times L = \pi \times 2 \times L$
Total length of the shaft = 10 m
Ignore the top 1.5 m of the shaft and the bottom "D" of the shaft for skin friction.
The diameter of the shaft is 2 m.

Effective length of the shaft = 10 − 1.5 − 2 = 6.5 m *(21.3 ft)*
$A_p = \pi \times d \times L = \pi \times 2 \times L = 40.8$ m²
c (cohesion) = 50 kPa
Skin friction = 0.5 × 50 × 40.8 kN
            = 1,020 kN   *(229 kips)*

STEP 3: Find the allowable caisson capacity.

$$P_{allowable} = S_u/FOS + Q_u/FOS - \text{Weight of the caisson} + \text{Weight of soil removed}$$

Weight of the caisson (W) = Volume of the caisson × Density of concrete
Assume density of concrete to be 23.5 kN/m³.
Weight of the caisson (W) = $(\pi \times D^2/4 \times L) \times 23.5$ kN
                = $(\pi \times 2^2/4 \times 10) \times 23.5$ kN
Weight of the caisson (W) = 738.3 kN   *(166 kips)*

Assume a factor of safety of 3.0 for skin friction and 2.0 for end bearing.

$P_{allowable} = S_u/FOS + Q_u/FOS - \text{Weight of the caisson} + \text{Weight of soil removed}$
$P_{allowable} = 1,413.8/3.0 + 1,020/2.0 - 738.3 = 243$ kN   *(55 kips)*

In this example, the weight of soil removed is ignored.

Note: $P_{allowable}$ obtained in this case is fairly low. Piles may be more suitable for this situation. If stiffer soil is available at a lower depth, the caisson can be placed at a much stronger soil. Another option is to consider a bell.

Design Example 7.2 (Caisson Design in Multiple Clay Layers)

Find the allowable capacity of a 1.5-m-diameter caisson placed at 15 m below the surface. The top layer was found to be clay with a cohesion of 60 kPa, and the bottom layer was found to have a cohesion of 75 kPa. The top clay layer has a thickness of 5 m. The caisson was constructed using drilling mud where it is not certain that the filter cake will be removed during concreting. The density of soil = 17 kN/m³ for both layers and groundwater is at 2 m below the surface.

Diameter of the cassion = 1.5 m

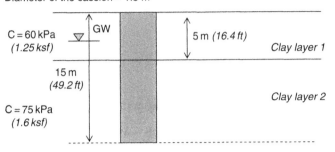

Figure 7.10 Caisson in a multiple clay layer

Solution

STEP 1: Find the end bearing capacity.

$$P_u = Q_u + S_u$$

End bearing capacity $(Q_u) = 9 \cdot c \cdot A$
End bearing capacity $(Q_u) = 9 \times 75 \times (\pi \cdot d^2)/4$

$$= 9 \times 75 \times (\pi \cdot 1.5^2)/4 = 1,192.8 \text{ kN} \quad (268 \text{ kips})$$

STEP 2: Find the skin friction in clay layer 1.

$$\text{Skin friction} = \alpha \cdot c \cdot A_p$$

$\alpha = 0.3$ (From Table 7.1. Assume that the filter cake will be not be removed.)

$$A_p = \pi \times d \times L = \pi \times 1.5 \times L$$

Ignore the top 1.5 m of the shaft and bottom "D" of the shaft for skin friction.
The effective length of the shaft in clay layer 1 = 5 – 1.5 = 3.5 m
$A_p = \pi \times d \times L = \pi \times 1.5 \times 3.5 = 16.5 \text{m}^2$
c (cohesion) = 60 kPa
Skin friction = $0.3 \times 60 \times 16.5$ kN
$$= 297 \text{ kN} \quad (67 \text{ kips})$$

STEP 3: Find the skin friction in clay layer 2.

$$\text{Skin friction} = \alpha \cdot c \cdot A_p$$

$\alpha = 0.3$ (From Table 7.1. Assume that the filter cake will be not be removed.)
$A_p = \pi \times d \times L = \pi \times 1.5 \times L$
Ignore the bottom "D" of the shaft.

Effective length of the shaft in clay layer $2 = 10 -$ Diameter $= 10 - 1.5$
$\quad = 8.5\,\text{m}$
$A_p = \pi \times d \times L = \pi \times 1.5 \times 8.5 = 40.1\,\text{m}^2$
c (cohesion) $= 75\,\text{kPa}$
Skin friction $= 0.3 \times 75 \times 40.1\,\text{kN}$
$\qquad\qquad\quad = 902.2\,\text{kN}$ *(203 kips)*

$\quad$ Total skin friction $= 297 + 902.2 = 1{,}199.2\,\text{kN}$

STEP 4: Find the allowable caisson capacity.

$$P_{\text{allowable}} = S_u/\text{FOS} + Q_u/\text{FOS} - \text{Weight of the caisson} \\ + \text{Weight of soil removed}$$

Weight of the caisson (W) $=$ Volume of the caisson $\times$ Density of
$\quad$ concrete
Assume the density of concrete to be $23.5\,\text{kN/m}^3$.
Weight of the caisson (W) $= (\pi \times D^2/4 \times L) \times 23.5\,\text{kN}$
$\qquad\qquad\qquad\qquad = (\pi \times 1.5^2/4 \times 15) \times 23.5\,\text{kN}$
Weight of the caisson (W) $= 622.9\,\text{kN}$

Assume a factor of safety of 3.0 for skin friction and 2.0 for end
bearing.

$P_{\text{allowable}} = S_u/\text{FOS} + Q_u/\text{FOS} - \text{Weight of the caisson}$
$P_{\text{allowable}} = 1{,}199.2/3.0 + 1{,}192.8/2.0 - 622.9 = 373.2\,\text{kN}$ *(84 kips)*

$\quad$ The weight of soil removed is ignored in this example.

7.4   Meyerhof Equation for Caissons

End Bearing Capacity

Meyerhof proposed the following equation based on the SPT (N) value
to compute the ultimate end bearing capacity of caissons. The Meyerhof equation was adopted by DM 7.2 as an alternative method to static
analysis.

$$q_{\text{ult}} = 0.13\,C_N \times N \times D/B$$

$q_{\text{ult}} =$ ultimate point resistance of caissons (tsf)
$\quad N =$ standard penetration resistance (blows/ft) near pile tip
$\quad C_N = 0.77 \log 20/p$
$\quad p =$ effective overburden stress at pile tip (tsf)

Note: "p" should be more than 500 psf. It is very rare for effective overburden stress at the pile tip to be less than 500 psf.

D = depth driven into granular (sandy) bearing stratum (ft)
B = width or diameter of the pile (ft)
q_1 = limiting point resistance (tsf), equal to 4 N for sand and 3 N for silt

Modified Meyerhof Equation

Meyerhof developed the above equation using many available load test data and obtaining average N values. Pile tip resistance is a function of the friction angle. For a given SPT (N) value, different friction angles are obtained for different soils.

For a given SPT (N) value, the friction angle for coarse sand is 7 to 8% higher than that for medium sand. At the same time, for a given SPT (N) value, the friction angle is 7 to 8% lower in fine sand compared to that for medium sand. Hence, the following modified equations are proposed:

$$q_{ult} = 0.15 \ C_N \times N \ D/B \ \text{tsf (coarse sand)}$$
$$q_{ult} = 0.13 \ C_N \times N \ D/B \ \text{tsf (medium sand)}$$
$$q_{ult} = 0.12 \ C_N \times N \ D/B \ \text{tsf (fine sand)}$$

Design Example 7.3

Find the tip resistance of the 4-ft diameter caisson shown using the modified Meyerhof equation. The SPT (N) value at caisson tip is 15 blows per foot.

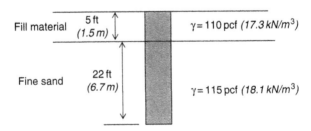

Figure 7.11   Caisson in a sand layer

Solution

Pile capacity comes form tip resistance and skin friction. In this example, only the tip resistance is calculated.

STEP 1: Ultimate point resistance for driven piles for fine sand =

$q_{tip} = 0.12 \ C_N \times N \times D/B$ tsf (fine sand)
$C_N = 0.77 \log 20/p$
  p = effective overburden stress at pile tip (tsf)
  $p = 5 \times 110 + 22 \times 115 = 3{,}080$ psf $= 1.54$ tsf   *(147 kPa)*
$C_N = 0.77 \log [20/1.54] = 0.86$
  D = depth into bearing stratum = 22 ft   *(6.7 m)*

Fill material is not considered to be a bearing stratum.

  B = 4 ft (width or diameter of the pile)
$q_{tip} = 0.12 \ C_N \times N \times D/B$ (fine sand)
$q_{tip} = 0.12 \times 0.86 \times 15 \times 22/4 = 8.5$ tsf   *(814 kPa)*

Maximum allowable point resistance = 4 N tsf for sandy soils

$$4 \times N = 4 \times 15 = 60 \text{ tsf}$$

Hence, $q_{tip} = 8.5$ tsf
Allowable point bearing capacity = 8.5/FOS
Assume a factor of safety of 3.0.

  Hence, the total allowable point bearing capacity $q_{tip \cdot \text{allowable\_tip}}$ = 2.84 tsf   *(272 kPa)*

$$Q_{tip \cdot \text{allowable}} \times \text{tip area} = q_{\text{allowable\_tip}} \times \pi \times (4^2)/4$$
$$= 36 \text{ tons}   (320 \text{ kN})$$

Only the tip resistance was computed in this example.

Meyerhof Equation for Skin Friction

Meyerhof proposes the following equation for skin friction for caissons:

  f = N/100 tsf
  f = unit skin friction (tsf)
  N = average SPT (N) value along the pile

Note: As per Meyerhof, unit skin friction "f" should not exceed 1 tsf. Here the Meyerhof equations have been modified to account for soil gradation.

f = N/92 tsf (coarse sand)
f = N/100 tsf (medium sand)
f = N/108 tsf (fine sand)

Design Example 7.4

Find the skin friction of the 4-ft-diameter caisson shown using the Meyerhof equation. Average SPT (N) value along the shaft is 15 blows per foot. Ignore the skin friction in fill material.

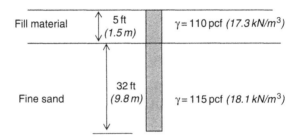

Figure 7.12   Caisson design using Meyerhof equation

Solution

Only the skin friction is calculated in this example.

STEP 1: For fine sand

Unit skin friction: f = N/108 tsf
Unit skin friction: f = 15/108 tsf = 0.14 tsf

Total skin friction = unit skin friction × perimeter surface area

Total skin friction = $0.14 \times \pi \times D \times L$
Total skin friction = $0.14 \times \pi \times 4 \times 32 = 56$ tons   *(498 kN)*
Allowable skin friction = 56/FOS

Assume a factor of safety (FOS) of 3.0

Allowable skin friction = 56/3.0 = 18.7 tons   *(166 kN)*

7.5   Caisson Design for Uplift Forces

Caissons are widely used to resist uplift forces. Uplift forces are caused by wind and buoyancy. Buoyancy could be a factor for buildings with basements where change in groundwater levels could create uplift forces.

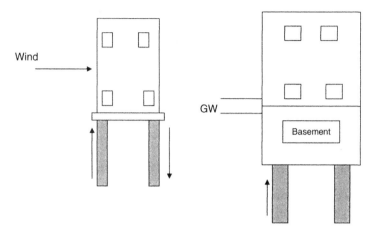

Figure 7.13   Uplift forces acting on caissons

Design Example 7.5

A telecommunication transmission tower is supported on a triangular foundation placed on three caissons as shown. Each caisson can be subjected to a tensile force of 700 kN due to wind loading acting on the tower. Each caisson should also be capable of providing a compressive capacity of 600 kN to account for the weight of the structure. The unconfined compressive strength (S_u) of the clay soil is found to be 100 kPa. Provide a suitable diameter and a length to account for above loading. The adhesion factor between soil and caisson (α) = 0.5 for tension and 0.6 for compression.

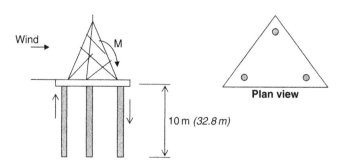

Figure 7.14   Telecommunication tower

STEP 1: Design the caissons for uplift forces.

Assume a diameter of 1.5 m and a length of 10 m for the caissons.

Allowable resistance against uplift force = Allowable skin friction + Weight of the caisson

The weight of the caisson itself will be helpful in resisting the uplift forces.

Ultimate skin friction = $(\alpha \cdot C) \times A_s$

C = cohesion = (unconfined compressive strength)/2
$= (Q_u)/2 = 100/2 \text{ kPa} = 50 \text{ kPa}$
α = adhesion factor = 0.5
A_s = surface area of the caisson = $(\pi \cdot D)$. length = $\pi \cdot 1.5 \cdot 10 = 47.1 \text{ m}^2$

Ultimate skin friction = $0.5 \times 50 \times (47.1) \text{ kN} = 1,177.5 \text{ kN}$
Allowable skin friction = $1177.5/F \cdot O \cdot S = 1177.5/3.0 = 392.5 \text{ kN}$ *(88.2 kips)*

A factor of safety of 3.0 is normally assigned for skin friction.

Weight of the caisson = (Volume of the caisson × Density of concrete)
$= (\pi \times D^2/4) \times \text{length} \times 23.5 \text{ (Density of concrete} = 23.5 \text{ kN/m}^3)$
$= (\pi \times 1.5^2/4) \times 10 \times 23.5 \text{ (Density of concrete} = 23.5 \text{ kN/m}^3)$
$= 415.3 \text{ kN} \quad (93.3 \text{ kips})$

Use a smaller factor of safety for the weight of the caisson since the density and volume of concrete is accurately known.

Allowable weight of the caisson = $415.3/\text{FOS} = 1,661/1.1 = 377.5 \text{ kN}$
(85 kips)
Allowable uplift resistance = 392.5 + 377.5 = 770 kN
Required uplift resistance = 700 kN *(157 kips)*
Caissons are sufficient to resist the uplift forces.

STEP 2: Design the caissons for compressive forces.

Allowable compressive capacity = Tip bearing resistance
+ Allowable skin friction − Weight of the caisson
+ Weight of soil removed

Tip bearing resistance = $N_c \cdot C \cdot A_t$

N_c = bearing capacity factor = 9
C = cohesion = 50 kN
A_t = area at the tip of the caisson = $(\pi \cdot D^2/4) = \pi \times (1.5)^2/4 = 1.767 \text{ m}^2$

Tip bearing resistance $= (9 \times 50 \times 1.767)\,\mathrm{kN} = 795.2\,\mathrm{kN}$
Allowable tip bearing resistance $= 795.2/2.0 = 397.6\,\mathrm{kN}$
Factor of safety of 2.0 is used for the end bearing resistance.
Ultimate skin friction $= (\alpha \cdot C) \times A_s$
"α" for compression is found to be 0.6.
Ultimate skin friction $= (0.6 \times 50) \times A_s$
Ultimate skin friction in compression $= (0.6 \times 50) \times (\pi \times 1.5 \times 10) = 1{,}413.7\,\mathrm{kN}$
Allowable skin friction in compression $= 1{,}413.7/3.0 = 471.2\,\mathrm{kN}$ *(106 kips)*
Allowable compressive capacity $=$ Allowable tip bearing resistance $+$
   Allowable skin friction − Weight of the caisson + Weight of soil removed
Allowable compressive capacity $= 397.6 + 471.2 - 415.3 = 453.5\,\mathrm{kN}$
   (102 kips)
The weight of soil removed is ignored in this example.
Required compressive capacity $= 600\,\mathrm{kN}$ *(135 kips)*

The assumed caisson diameter and length are not sufficient to provide adequate compressive capacity.
The following solutions are available for the designer.

a. Increase the diameter or the length of the caisson.

b. Provide a bell at the bottom.

c. Increase piles to a deeper depth.

d. Create a large shallow foundation.

7.6   Caisson Design in Sandy Soils

As in piles, the capacity of caissons comes from the end bearing and the skin friction.

Ultimate caisson capacity $(P_u) =$ Ultimate end bearing capacity (Q_u)
$+$ Ultimate skin friction (S_u)

$$P_u = Q_u + S_u$$

End bearing capacity $(Q_u) = q_p/\alpha \times A_p$

q_p is obtained from the following table.

Table 7.3   "q_p" values for different sand conditions

Sand State at the Bottom of the Caisson	kPa	ksf
Loose sand (not recommended)	0	0
Medium-dense sand	1,600	32
Dense sand	4,000	08

(*Source*: Reese et al., 1976)

$\alpha = 2.0\,B$ for base width "B" in meters
$\alpha = 0.6\,B$ for base width "B" in feet
Skin friction $= K \cdot p_v \cdot \tan \delta \cdot (A_p)$
$K =$ lateral earth pressure coefficient

Table 7.4   K Value

Depth to Base (m)	K
Less than or equal to 7.5 m	0.7
Between 7.5 m and 12 m	0.6
Greater than 12 m	0.5

(*Source*: Reese et al., 1976)

$p_v =$ stress in the perimeter of the caisson. Usually taken as the soil pressure at midpoint of the caisson.
$\delta = \varphi$; for caissons in sandy soils
$A_p =$ perimeter surface area of the caisson

Allowable Caisson Capacity

Allowable capacity of the caisson is given by;

$$P_{allowable} = S_u + Q_u/FOS - \text{Weight of the caisson} + \text{Weight of soil removed}$$

Only the end bearing capacity is divided by a factor of safety. The weight of the caisson should be deducted to obtain the allowable column load.

AASHTO Method for End Bearing Capacity

The AASHTO Method

The ultimate end bearing capacity in caissons, placed in sandy soils, is given by:

$$Q_u = q_t \cdot A$$
$$q_t = 1.20 \, N \, ksf \quad (0 < N < 75)$$

N = standard penetration test value (blows/ft)
Metric units $q_t = 57.5 \, N \, kPa \ (0 < N < 75)$
A = cross-sectional area at the bottom of the shaft
For all N values above 75; $q_t = 90$ ksf (in psf units)
$q_t = 4{,}310$ kPa (in metric units)

Design Example 7.6 (Caisson Design in Sandy Soil)

Find the ultimate capacity of a 2-m-diameter caisson placed at 10 m below the surface. The soil was found to be medium dense. The density of soil = $17 \, kN/m^3$, and the friction angle of the soil is 30°.

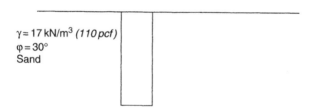

Figure 7.15   Caisson in sandy soils

Solution

$$P_u = Q_u + S_u$$

End bearing capacity $(Q_u) = (q_p/\alpha) \times A$
Skin friction $(S_u) = K \cdot p_v \cdot \tan \delta \cdot (A_p)$

STEP 1: Find the end bearing capacity.

$$\text{End bearing capacity } (Q_u) = (q_p/\alpha) \times A_p$$

$q_p = 1{,}600$ kPa from Table 7.3 for medium-dense sand
$\alpha = 2.0\,B$ for SI units $= 2.0 \times 2 = 4$

A = cross-sectional area of the caisson = $(\pi \cdot D^2/4) \cdot L = (\pi \cdot 2^2/4)$
    = $3.14 \, m^2$
End bearing capacity (Q_u) = $(1,600/4) \times 3.14 \, kN$
    = $1,256.6 \, kN$ (282 kips)

STEP 2: Find the skin friction.

$$Skin \; friction = K \cdot p_v \cdot tan\delta \cdot (A_p)$$

$K = 0.6$ (From Table 7.4 provided by Reese. The depth to base is 10 m.)
p_v = effective stress at the midpoint of the caisson
$p_v = 17 \times 5.0 = 85 \, kN/m^2$
$\delta = \varphi$ for sandy soils; hence, $\delta = 30°$
$A_p = (\pi \cdot d) \times L = (\pi \cdot 2) \times 10 = 62.8 \, m^2$
Ultimate skin friction = $K \cdot p_v \cdot tan \, \delta \cdot (A_p)$
Ultimate skin friction (S_u) = $0.6 \times 85 \times (tan \; 30) \times 62.8$
    = $1,849 \, kN$ (416 kips)

STEP 3: Find the allowable caisson capacity.

$$P_{allowable} = S_u + Q_u/FOS - Weight \; of \; the \; caisson$$
$$+ \; Weight \; of \; soil \; removed$$

Weight of the caisson (W) = Volume of the caisson $\times$ Density of
    concrete
Assume density of concrete to be $23.5 \, kN/m^3$
Weight of the caisson (W) = $(\pi \times D^2/4 \times L) \times 23.5 \, kN$
    = $(\pi \times 2^2/4 \times L) \times 23.5 \, kN$
Weight of the caisson (W) = $738.3 \, kN$
Assume a factor of safety of 3.0.
$P_{allowable}$ = $S_u + Q_u/FOS$ - Weight of the caisson + Weight of soil
    removed
$P_{allowable}$ = $1,849 + 1,256.6/3.0 - 738.3 = 1,530 \, kN$ (344 kips)

The weight of soil removed is ignored in this example.

Design Example 7.7 (Caisson Design in Sandy Soil—Multiple Sand Layers)

Find the ultimate capacity of a 2-m-diameter caisson placed at 12 m below the surface. Soil strata are as shown in figure 7.16. Groundwater is 2 m below the surface. Find the allowable capacity of the caisson.

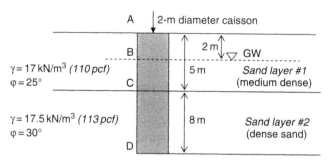

Figure 7.16   Caisson in a multiple sand layers

Solution

$$P_u = Q_u + S_u$$

End bearing capacity $(Q_u) = (q_p/\alpha) \times A$
Skin friction $(S_u) = K \cdot p_v \cdot \tan \delta \cdot (A_p)$

STEP 1: Find the end bearing capacity.

End bearing capacity $(Q_u) = (q_p/\alpha) \times A$

$q_p = 4,000$ kPa from Table 7.3 for dense sand
$\alpha = 2.0\,B$ for SI units $= 2.0 \times 2 = 4$
$A =$ cross-sectional area of the caisson $= (\pi \cdot D^2/4) \cdot L = (\pi \cdot 2^2/4)$
$\quad = 3.14\,\mathrm{m}^2$
End bearing capacity $(Q_u) = (4,000/4) \times 3.14\,\mathrm{kN} = 3,140\,\mathrm{kN}$ *(706 kips)*

STEP 2: Find the skin friction in sand layer #1 (above groundwater)
(A to B).

$$\text{Skin friction} = K \cdot p_v \cdot \tan\delta \cdot (A_p)$$

$K = 0.5$ (The depth to base is 13 m; hence, from Table 7.4, $K = 0.5$.)
$p_v =$ effective stress at the midpoint of the sand layer 1 from A to B
$p_v = 17 \times 1 = 17\,\mathrm{kN/m}^2$
$\delta = \varphi$ for sandy soils; hence, $\delta = 25°$ for sand layer 1
$A_p = (\pi \cdot d) \times L = (\pi \cdot 2) \times 2 = 12.6\,\mathrm{m}^2$
Ultimate skin friction $= K \cdot p_v \cdot \tan\delta \cdot (A_p)$
Ultimate skin friction (S_u) (A to B) $= 0.5 \times 17 \times (\tan 25) \times 12.6$
$\quad = 49.9\,\mathrm{kN}$

STEP 3: Find the skin friction in sand layer #1 (below groundwater) (B to C).

$$\text{Skin friction} = K \cdot p_v \cdot \tan\delta \cdot (A_p)$$

K = 0.5 (The depth to base is 13 m; hence, from Table 7.4, K = 0.5.)
p_v = effective stress at the midpoint of the sand layer 1 from B to C
$p_v = (\gamma \times 2 + (\gamma - \gamma_w) \times 1.5 = 17\,\text{kN/m}^2$
$p_v = 17 \times 2 + (17 - 9.8) \times 1.5 = 44.8\,\text{kN/m}^2$
$\delta = \varphi$ for sandy soils; hence, $\delta = 25°$ for sand layer 1
$A_p = (\pi \cdot d) \times L = (\pi \cdot 2) \times 2 = 12.6\,\text{m}^2$
Ultimate skin friction $= K \cdot p_v \cdot \tan\delta \cdot (A_p)$
Ultimate skin friction (S_u) (B to C) $= 0.5 \times 44.8 \times (\tan 25) \times 12.6$
  $= 131.6\,\text{kN}$ *(29.7 kips)*

STEP 4: Find the skin friction in sand layer #2 (below groundwater) (C to D).

$$\text{Skin friction} = K \cdot p_v \cdot \tan\delta \cdot (A_p)$$

K = 0.5 (The depth to base is 13 m; hence from 7.4, K = 0.5.)
p_v = effective stress at the midpoint of the sand layer 2 from C to D
$p_v = (\gamma_1 \times 2 + (\gamma_1 - \gamma_w) \times 3 + (\gamma_2 - \gamma_w) \times 4 = 17\,\text{kN/m}^2$
$p_v = 17 \times 2 + (17 - 9.8) \times 3 + (17.5 - 9.8) \times 4 = 86.4\,\text{kN/m}^2$ *(1,804 psf)*
$\delta = \varphi$ for sandy soils; hence, $\delta = 30°$ for sand layer 2
$A_p = (\pi \cdot d) \times L = (\pi \cdot 2) \times 8 = 50.3\,\text{m}^2$
Ultimate skin friction $= K \cdot p_v \cdot \tan\delta \cdot (A_p)$
Ultimate skin friction (Su) (C to D) $= 0.5 \times 86.4 \times (\tan 30) \times 50.3$
  $= 1,254\,\text{kN}$

Total skin friction:

Skin friction from A to B = 49.9
Skin friction from B to C = 131.6
Skin friction from C to D = 1,254
Total skin friction from A to D = 49.9 + 131.6 + 1,254
  = 1,436 kN   *(322 kips)*

STEP 5: Find the allowable caisson capacity.

$$P_{\text{allowable}} = S_u + Q_u/\text{FOS} - \text{Weight of the caisson} + \text{Weight of soil removed}$$

Weight of the caisson (W) = Volume of the caisson × Density of concrete
Assume density of concrete to be $23.5 \, kN/m^3$
Weight of the caisson (W) = $(\pi \times D^2/4 \times L) \times 23.5 \, kN$
$$= (\pi \times 2^2/4 \times 13) \times 23.5 \, kN$$
Weight of the caisson (W) = 959.8 kN   *(216 kips)*
Assume a factor of safety of 3.0 for skin friction and 2.0 for end
    bearing.
$P_{allowable}$ = S_u/FOS + Q_u/FOS − Weight of the caisson
$P_{allowable}$ = 1,436/3.0 + 3,140/2.0 − 959.8 = 1,088 kN   *(244 kips)*

Weight of soil removed was ignored in this example.

7.7   Belled Caisson Design

Belled caissons are used to increase the end bearing capacity. Unfortunately, one loses the skin friction in the bell area since experiments
have shown the skin friction in the bell to be negligible (Reese 1976).
In addition, Reese and O'Neill (1988) and ASHTO suggest excluding
the skin friction in a length equal to the shaft diameter above the bell.
(See Figure 7.17.)

    Belled caissons are usually placed in clay soils. It is almost impossi-
ble to create a bell in sandy soils. On rare occasions, with special
equipment belled caissons are constructed in sandy soils.

    Belled caissons are used to increase the end bearing capacity. Unfor-
tunately, one loses the skin friction in the bell area and one diameter
above the bell (Reese and O'Neill, 1988).

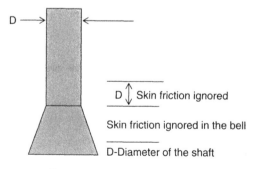

Figure 7.17   Belled caisson

Skin friction in the bell and one diameter of the shaft (D) above the bell are ignored.

The ultimate capacity of belled caissons is given by the following equation.

Ultimate capacity of the belled caisson = Ultimate end bearing
capacity + Ultimate skin friction

$P_u = Q_u + S_u$

$Q_u = 9 \cdot c \cdot$ Bottom cross-sectional area of the bell

$c =$ cohesion

Area of the bell $= \pi \times d_b^2/4$ ($d_b =$ bottom diameter of the bell)

$S_u = \alpha \times c \times$ perimeter surface area of the shaft (ignore the bell)

$S_u = \alpha \times c \times (\pi \times d \times L)$

$d =$ diameter of the shaft

$L =$ length of the shaft portion.

$\alpha = 0.55$ for all soil conditions unless found by experiments (AASHTO)

Design Example 7.8

Find the allowable capacity of the belled caisson shown. The diameter of the bottom of the bell is 4 m, and the height of the bell is 2 m. The diameter of the shaft is 1.8 m, and the height of the shaft is 10 m. Cohesion of the clay layer is 100 kN/m². Adhesion factor (α) was found to be 0.55. Ignore the skin friction in the bell and one diameter of the shaft above the bell.

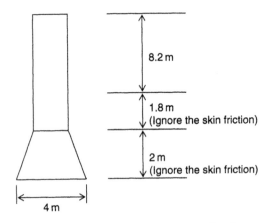

8.2 m

1.8 m
(Ignore the skin friction)

2 m
(Ignore the skin friction)

4 m

Figure 7.18   Belled caisson in a clay layer

Solution

STEP 1: Ultimate caisson capacity $(P_u) = Q_u + S_u - W$

$$
\begin{aligned}
Q_u &= \text{ultimate end bearing capacity} \\
&= 9 \times c \times (\text{area of the bottom of the bell}) \\
&= 9 \times 100 \times (\pi \times 4^2/4) = 11{,}310\,\text{kN} \quad (2{,}543\ kips)
\end{aligned}
$$

S_u = ultimate skin friction
W = Weight of the caisson

STEP 2: Find the ultimate skin friction.

$$
\begin{aligned}
\text{Ultimate skin friction } (S_u) &= \alpha \times c \times (\pi \times d \times L) \\
\text{Ultimate skin friction } (S_u) &= 0.55 \times 100 \times (\pi \times 1.8 \times 8.2) \\
&= 2550\ \text{kN} \quad (573\ kips)
\end{aligned}
$$

The height of the shaft is 10 m, and length equal to one diameter of the shaft above the bell is ignored. Hence, the effective length of the shaft is 8.2 m.

STEP 3: Find the weight of the caisson.

Assume the density of concrete to be $23\,\text{kN/m}^3$.
Weight of the shaft $= (\pi \times d^2/4) \times 10 \times 23$
$= (\pi \times 1.8^2/4) \times 10 \times 23\,\text{kN} = 585.3\,\text{kN} \quad (131\ kips)$
Find the weight of the bell.
Average diameter of the bell $(d_a) = (1.8 + 4)/2 = 2.9\,\text{m}$
Use the average diameter of the bell (d_a) to find the volume of the bell.
Volume of the bell $= \pi \times d_a{}^2/4 \times h$
h = height of the bell.
$\pi \times 2.9^2/4 \times 2 = 13.21\,\text{m}^3$
Weight of the bell = Volume × density of concrete $= 13.21 \times 23$
$= 303.8\,\text{kN} \quad (68.3\ kips)$

STEP 4:

$$
\begin{aligned}
\text{Ultimate caisson capacity}(P_u) = Q_u + S_u &- \text{Weight of the caisson} \\
&+ \text{Weight of soil removed}
\end{aligned}
$$

Allowable caisson capacity $= 11{,}310/\text{FOS} + 2318/\text{FOS} - 585.3 - 303.8$

Assume a factor of safety of 2.0 for end bearing and 3.0 for skin friction. Since the weight of the caisson is known fairly accurately, no safety factor is needed. The weight of soil removed is ignored in this example.

Allowable caisson capacity $= 11{,}310/2.0 + 2550/3.0 - 585.3 - 303.8$

Allowable caisson capacity $= 5{,}615\,kN$ *(1,262 kips)*

Design Example 7.9

Find the allowable capacity of the shaft in the previous example, assuming a bell was not constructed. Assume the skin friction is mobilized in the full length of the shaft.

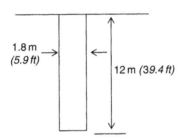

Figure 7.19   Caisson in a clay layer

Solution

STEP 1: Ultimate end bearing capacity $(Q_u) = 9 \times c \times (\pi \times 1.8^2/4)$

Since there is no bell, the new diameter at the bottom is 1.8 m.
Ultimate end bearing capacity $(Q_u) = 9 \times 100 \times (\pi \times 1.8^2/4)$
   $= 2{,}290\,kN$ *(515 kips)*

STEP 2: Ultimate skin friction $(S_u) = \alpha \times c \times (\pi \times 1.8) \times 12$

The new height of the shaft is 12 m since a bell is not constructed.

Ultimate skin friction $(S_u) = 0.55 \times 100 \times (\pi \times 1.8) \times 12$
                $= 3{,}732\,kN$ *(839 kips)*

Note: It is assumed that the skin friction of the shaft is mobilized along the full length of the shaft for 12 m.

STEP 3: Find the weight of the shaft.

W = Weight of the shaft = $(\pi \times d^2/4) \times L \times$ Density of concrete

W = Weight of the shaft = $(\pi \times 1.8^2/4) \times 12 \times 23 = 702.3$ kN (*158 kips*)

STEP 4: Find the allowable caisson capacity.

Allowable caisson capacity = $Q_u/FOS + S_u/FOS -$ Weight of caisson + Weight of soil removed

Assume a factor of safety of 2.0 for end bearing and 3.0 for skin friction. Since the weight of the caisson is known fairly accurately, no safety factor is needed.

$$\text{Allowable caisson capacity} = 2{,}290/2.0 + 3{,}732/3.0 - 702.3 \text{ kN}$$
$$= 1{,}686.7 \text{ kN}$$

Note: The weight of soil removed is ignored in this example.

In the previous example, we found the allowable capacity with the bell to be 5,615 kN, significantly higher than the straight shaft.

Design Example 7.10

Find the allowable capacity of the belled caisson shown. The diameter of the bottom of the bell is 4 m, and the height of the bell is 2 m. The diameter of the shaft is 1.8 m, and the height of the shaft is 11.8 m. The cohesion of the clay layer is 100 kN/m². The adhesion factor (α) was found to be 0.5. Ignore the skin friction in the bell and one diameter above the bell.

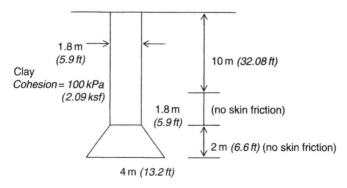

Figure 7.20   Belled caisson example in clay soil

Solution

STEP 1: Ultimate caisson capacity $(P_u) = Q_u + S_u -$ Weight of caisson + Weight of soil removed

$$Q_u = \text{ultimate end bearing capacity}$$
$$= 9 \times c \times \text{(area of the bottom of the bell)}$$
$$= 9 \times 100 \times (\pi \times 4^2/4) = 11{,}310\,\text{kN} \quad (2{,}542\ kips)$$
$$S_u = \text{ultimate skin friction}$$

STEP 2: Find the ultimate skin friction.

$$\text{Ultimate skin friction } (S_u) = \alpha \times c \times (\pi \times d \times L)$$
$$\text{Ultimate skin friction } (S_u) = 0.5 \times 100 \times (\pi \times 1.8 \times 10)$$
$$= 2827\,kN \quad (636\ kips)$$

Skin friction in 1.8 m (length equal to diameter of the shaft) is ignored.

STEP 3: Find the weight of the caisson.

Assume the density of concrete to be $23\,\text{kN/m}^3$.

$$\text{Weight of the shaft} = (\pi \times d^2/4) \times 11.8 \times 23$$
$$= (\pi \times 1.8^2/4) \times 11.8 \times 23\,\text{kN}$$
$$= 690.6\,\text{kN} \quad (155\ kips)$$

Find the weight of the bell.

Average diameter of the bell $(d_a) = (1.8 + 4)/2 = 2.9\,\text{m} \quad (9.5\ ft)$

Use the average diameter of the bell (d_a) to find the volume of the bell.

Volume of the bell $= \pi \times d_a^2/4 \times h$
h = height of the bell.
$\pi \times 2.9^2/4 \times 2 = 13.21\,\text{m}^3$
Weight of the bell $=$ Volume $\times$ density of concrete $= 13.21 \times 23$
$$= 303.8\,\text{kN}$$

STEP 4: Ultimate caisson capacity $(P_u) = Q_u + S_u -$ Weight of caisson + Weight of soil removed

Allowable caisson capacity $= 11{,}310/\text{FOS} + 2827/\text{FOS} - 690.6 - 303.8$

Note: The weight of soil removed ignored in this example.

Assume a factor of safety of 2.0 for end bearing and 3.0 for skin friction. Since the weight of the caisson is known fairly accurately, no safety factor is needed.

Allowable caisson capacity $= 11,310/2.0 + 2827/3.0 - 690.6 - 303.8$
Allowable caisson capacity $= 5,602$ kN   *(1,259 kips)*

7.8   Settlement of Caissons

Total settlement of a caisson is given by the following equation for caissons subjected to compressive loads as shown.

$$S_{total} = S_a + S_{tip} + S_{skin}$$

S_a = settlement of the caisson due to axial deformation
S_{tip} = settlement due to tip load
S_{skin} = settlement due to skin friction loading

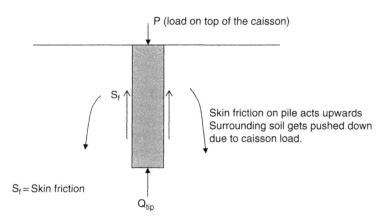

Figure 7.21   Development of skin friction

When a caisson is loaded, some of the load will be taken by skin friction and the rest will be transferred to the bottom (tip) of the caisson.

P = load applied at the top of the caisson
Q_{tip} = load transferred to the tip of the caisson
$P = Q_{tip} + S_f$ (skin friction)

Axial Deformation (S_a)

Owing to the load on the caisson, the caisson material will undergo deformation. If there is no skin friction, this deformation can be found using Young's elastic equation.

$$e = FL/AE$$

e = deformation;
F = force;
L = length of the caisson;
A = cross-sectional area
E = Young's modulus

   Unfortunately, this equation cannot be used when skin friction is present. "F" (force) acting along the caisson is not a constant in the presence of skin friction.

Deformation Due to Tip Load (S_{tip})

Settlement would occur due to loading at the tip of the caisson.

Deformation Due to Skin Friction (S_{skin})

Skin friction on the caisson acts upward. All actions have an equal and opposite reaction. If the skin friction is acting upward on the caisson, it should act downward on the surrounding soil. (See Figure 7.21.) Settlement would occur due to downward loading on the surrounding soil.

Methodology to Compute Axial Deformation (S_a)

Methodology proposed by Kulhawy (1990) is provided here.
   Axial deformation (S_a) is given by

$$S_a = (Q_{tip} + 2/3\ S_f) \times D/(AE_c)$$

Q_{tip} = load transferred to the tip of the caisson
   S_f = skin friction

Note that $P = Q_{tip} + S_f$; P = load acting on top of the caisson

D = depth of the caisson
A = cross-sectional area of the caisson
E_c = Young's modulus of concrete

The following table can be used to obtain the Q_{tip} value.

Table 7.5   Foundation stiffness factor

D/B	K_{fs} (foundation stiffness factor)	Q_{tip}/P
D = Depth of the caisson B = Diameter of the caisson	$K_{fs} = E_{concrete}/E_{soil}$	
2	100	0.20
	1,000	0.25
	10,000 and over	0.3
5	100	0.15
	1,000	0.17
	10,000 and over	0.18
10	100	0.09
	1,000	0.095
	10,000 and over	0.1
25	100	0.03
	1,000	0.04
	10,000 and over	0.05
50	100	0.01
	1,000	0.02
	10,000 and over	0.03

Source: Fang, H., and Kulhawy, F., Foundation Engineering Handbook, 1990.

First obtain the D/B value for the caisson. (D = depth of the caisson, B = diameter of the caisson)
Find K_{fs} (foundation stiffness factor).
$K_{fs} = E_{concrete}/E_{soil}$
$E_{concrete}$ = Young's modulus of concrete is approximately 4.8×10^5 ksf. (23×10^9 N/m^2).
This value depends on the concrete mix and curing process.
E_{soil} = Young's modulus of soil

The following table provides the range of E value for various soil types.

Table 7.6 Young's Modulus

Soil Type	E (Young's Modulus) Tons/ft^2	MN/m^2
Loose sand	50–200	5–20
Medium sand	200–500	20–50
Dense sand	500–1000	50–100
Soft clay	25–150	2.5–15
Medium clay	150–500	15–50
Stiff clay	500–2000	50–200

Source: Fang, H., Kulhawy, F. Foundation Engineering Handbook, 1990

Methodology to Compute Settlement Due to Tip Load (S_{tip})

Settlement due to tip load is given by

$$S_{tip} = C_t \times Q_{tip}/(B \times q_{ult})$$

S_{tip} = settlement due to tip load
C_t = empirical parameter
Q_{tip} = load transferred to the tip of the caisson

The following table is provided to find C_t.

Table 7.7 Sand Consistency vs. C_t

Soil Type	C_t (for Caissons)
Sand (loose to dense)	0.09–0.18
Clay (stiff to soft)	0.03–0.06
Silt (dense to loose)	0.09–0.12

Source: Fang, H., and Kulhawy, F., Foundation Engineering Handbook, 1990.

B = diameter of the caisson
q_{ult} = ultimate bearing capacity of the caisson
$q_{ult} = 9 \cdot c$ for clayey soils (c = cohesion)

Methodology to Compute Settlement Due to Skin Friction (S_{skin})

Settlement due to skin friction is given by

$$S_{skin} = C_s \times S_f/(D \times q_{ult})$$

S_{skin} = settlement due to skin friction
  C_s = empirical parameter
  $Cs = [0.93 + 0.16 \times (D/B)^{1/2}] \times C_t$
  D = depth of the caisson; B = diameter; C_t = empirical parameter
  q_{ult} = ultimate bearing capacity

Design Example 7.11

Find the total (long-term and short-term) settlement of the caisson shown. Soil is found to be medium stiff clay.

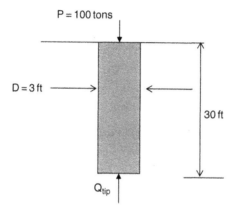

Figure 7.22   Caisson settlement

STEP 1: Total settlement $S_{total} = S_a + S_{tip} + S_{skin}$

  S_a = settlement of the caisson due to axial deformation
  S_{tip} = settlement due to tip load
S_{skin} = settlement due to skin friction loading

STEP 2: Compute the axial deformation (S_a).

$$S_a = (Q_{tip} + 2/3 \; S_f) \times D/(AE_c)$$

  Q_{tip} = load transferred to the tip of the caisson
    S_f = skin friction
    D = depth of the caisson = 30 ft
    A = cross-sectional area of the caisson = $\pi \times 3^2/4 = 7.07 \, ft^2$
$E_{concrete}$ = Young's modulus of concrete = 4.8×10^5 ksf
    E_{soil} = 150 to 500 tsf for medium clay (15–50 MN/m^2)

Assume 250 tsf for E_{soil}.

$K_{fs} = E_{concrete}/E_{soil} = 4.8 \times 10^5$ ksf/250 tsf $= 4.8 \times 10^5$ ksf/500
ksf $= 960$

$D/B = 30/3 = 10$

From Table 7.5; $\quad Q_{tip}/P = 0.095 \quad\quad P = 100$ tons (given)

Hence, $Q_{tip} = 100 \times 0.095 = 9.5$ tons $= 19$ kips

$S_f = P - Q_{tip} = 100 - 9.5 = 90.5$ tons $= 181$ kips

$S_a = (Q_{tip} + 2/3\, S_f) \times D/(AE_c)$

$S_a = (19 + 2/3 \times 181) \times 30/(7.07 \times 4.8 \times 10^5)$ ft $= 0.00123$ ft $= 0.015$ in.

STEP 3: Compute the settlement due to tip load (S_{tip}).

Settlement due to tip load is given by

$$S_{tip} = C_t \times Q_{tip}/(B \times q_{ult})$$

S_{tip} = Settlement at due to tip load

C_t = empirical parameter (obtained from Table 7.7)
From the table for stiff to soft clay $C_t = 0.03$ to 0.06.
Assume $C_t = 0.05$.
 B = diameter of the caisson
q_{ult} = ultimate bearing capacity of the caisson
$q_{ult} = 9 \cdot c$ for clayey soils (c = cohesion)
$q_{ult} = 9 \times 1,000 = 9,000$ psf $= 9$ ksf
$S_{tip} = C_t \times Q_{tip}/(B \times q_{ult})$
$S_{tip} = 0.05 \times 19/(3 \times 9)$ ft $= 0.035$ ft $= 0.42$ in.

Compute the Settlement due to Skin Friction (S_{skin})

Settlement due to skin friction is given by

$$S_{skin} = C_s \times S_f/(D \times q_{ult})$$

S_{skin} = settlement due to skin friction
 C_s = empirical parameter
 $C_s = [0.93 + 0.16 \times (D/B)^{1/2}] \times C_t$
 D = depth of the caisson; B = diameter; Ct = Empirical parameter
 (Table 7.7)
q_{ult} = Ultimate bearing capacity
 $C_s = [0.93 + 0.16 \times (30/3)^{1/2}] \times 0.05 = 0.072$

$$S_{skin} = C_s \times S_f/(D \times q_{ult})$$

$S_f = 181$ kips (calculated in step 2)

$S_{skin} = 0.072 \times 181/(10 \times 9)$ ft $= 0.145$ ft $= 1.73$ in.

Summary

S_a = settlement due to axial deformation = 0.015 in.

S_{tip} = settlement due to tip load = 0.42 in.

S_{skin} = settlement due to skin friction loading = 1.73 in.

Total settlement (long-term and short-term) = 2.17 in.   *(5.5 cm)*

This value is excessive, configuration of the caisson needs to be changed, and settlement should be computed again. The largest settlement occurs due to skin friction loading in surrounding soil. Increasing the depth and reducing the diameter provides one solution to the problem.

Reference

Fang, H., *Foundation Engineering Handbook*, Kulhawy, F, "Drilled shaft foundations," in Springer, 1990.

8

Design of Pile Groups

8.1   Introduction

Typically, piles are installed in a group and provided with a pile cap. The column is placed on the pile cap so that the column load is equally distributed among the individual piles in the group.

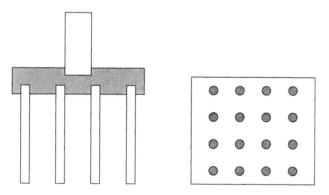

Figure 8.1   Pile groups

The capacity of a pile group is obtained by using an efficiency factor.

$$\text{Pile group capacity} = \text{Efficiency of the pile group} \times \text{Single pile capacity} \times \text{Number of piles}$$

If the pile group contains 16 piles and capacity of a single pile is 30 tons and the group efficiency is found to be 0.9, the group capacity is 432 tons.

Pile group capacity $= 0.9 \times 30 \times 16 = 432$ tons

Clearly, high pile group efficiency is desirable; the question is how to improve the group efficiency. Pile group efficiency is dependent on the spacing between piles. When the piles in the group are closer together, the pile group efficiency decreases. When the piles are placed far apart, efficiency increases, but because the size of the pile cap has to be larger, the cost of the pile cap increases.

Soil Disturbance during Driving

What happens in a pile group? When piles are driven, the soil surrounding the pile is disturbed. Disturbed soil has less strength than undisturbed soil. Some of the piles in the group are installed in partially disturbed soil causing them to have less capacity than others. Typically, piles in the center are driven first.

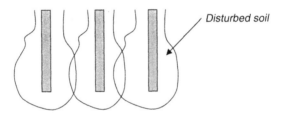

Figure 8.2   Soil disturbance due to pile driving

Soil disturbance caused by one pile impacts the capacity of adjacent piles. The group efficiency can be improved by placing the piles at a larger spacing. In clay soils, shear strength is reduced owing to disturbance.

Soil Compaction in Sandy Soil

When driving piles in sandy soils, surrounding soil will be compacted. Compacted soil tends to increase the skin friction of piles. Pile group placed in sandy soils may have a larger than one group efficiency. Soil compaction due to pile driving will be minimal in clay soils.

Pile Bending

When driving piles in a group, some piles can be bent owing to soil movement. This effect is more pronounced in clayey soils.

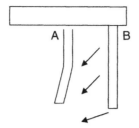

Figure 8.3   Pile bending

Assume that pile A is driven first and pile B is driven next. Soil movement caused by pile B can bend pile A as shown. This in return creates a lower group capacity.

End Bearing Piles

Piles that mainly rely on end bearing capacity may not be affected by other piles in the group.

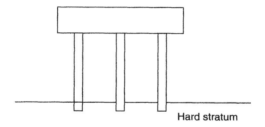

Figure 8.4   End bearing piles

Piles ending in very strong bearing stratum or in rock: When piles are not dependent on skin friction, group efficiency of 1.0 can be used.

Various guidelines for computing group capacity are given next.

AASHTO (1992) Guidelines

The American Association of State Highway and Transportation (AASHTO) provides the following guidelines.

AASHTO considers three situations:

a. Pile group in cohesive soils (clays and clayey silts)

b. Pile group in noncohesive soils (sands and silts)

c. Pile group in strong soil overlying weaker soil

The following tables use AASHTO guidelines for pile group efficiency in cohesive soils.

Pile Group Efficiency for Clayey Soils

Pile Spacing (center to center)	Group Efficiency
3 D	0.67
4 D	0.78
5 D	0.89
6 D or more	1.00
D = Diameter of piles	

Pile Group Efficiency for Sandy Soils

Pile Spacing (center to center)	Group Efficiency
3 D	0.67
4 D	0.74
5 D	0.80
6 D	0.87
7 D	0.93
8 D or more	1.00
D = Diameter of piles	

Design Example: Pile Group Capacity

The pile group is constructed in clayey soils as shown in Figure 8.5. Single-pile capacity was computed to be 30 tons, and each pile is 12 in. in diameter. The center-to-center distance of piles is 48 in. Find the capacity of the pile group using the AASHTO method.

Figure 8.5   Pile group – plan view

Solution

Pile group capacity = Efficiency of the pile group × Single pile
capacity × number of piles

Center-to-center distance between piles = 48 in.
Since the diameter of piles is 12 in., center to center distance is 4 D.
Efficiency of the pile group = 0.78 (AASHTO table)
Pile group capacity = 0.78 × 30 × 4 = 93.6 tons

Pile Group Capacity When Strong Soil Overlies Weaker Soil: (AASHTO)

Usually, piles are ended in strong soils. In some cases, a weaker
soil stratum may lie underneath the strong soil strata. In such situations, settlement due to weaker soil underneath has to be computed.

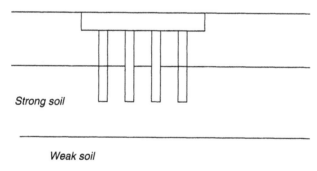

Strong soil

Weak soil

Figure 8.6   Friction piles

Pile Spacing (Center-to-Center Distance): International Building Code Guidelines

In no case should minimum distance be less than 24 in.

For circular piles: minimum distance (center-to-center)—twice the
average diameter of the butt.
Rectangular piles (minimum center-to-center distance)—3/4 times the
diagonal for rectangular piles.

Tapered piles (minimum center-to-center distance)—Twice the diameter at 1/3 of the distance of the pile measured from top of pile.

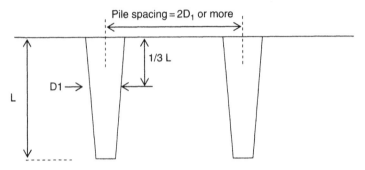

Figure 8.7 Tapered piles

8.2 Eccentric Loading on a Pile Group

Usually, piles are arranged in a group, so that the load will be equally distributed and no additional loads will be generated due to bending moments.

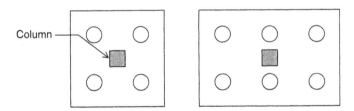

Figure 8.8 Eccentric loading on a pile group

When a pile group is subjected to an eccentric load, individual piles have different loads. The following example shows how to calculate the load on individual piles.

Design Example 1

A 10-ton column load is acting on point K. Find the loads on piles 1, 2, 3, and 4.

$$\text{Eccentricity (e)} = 1 \text{ ft}$$

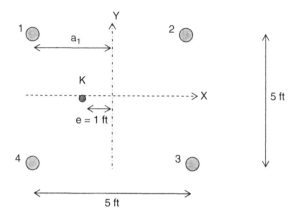

Figure 8.9   Eccentric loading

Solution

STEP 1: If the column load were to act on the center of gravity, all four columns would get 2.5 tons each.

- Since the load is acting at point K, the load on piles 1 and 4 would be larger and the load on piles 2 and 3 would be smaller.

- Assume loads on piles to be R_1, R_2, R_3 and R_4 respectively. The area of a pile is considered to be A.

Stress of a pile is given by Eq. (1). The first term $C/(n \cdot A)$ represents the load on each pile if there is no eccentricity. The second term $(C \cdot e \cdot a_1)/I$ represents the stress due to bending moment induced by eccentricity.

The stress on some piles increases due to eccentricity, while stress on some other piles decreases. For example, stress on piles 1 and 4 would increase due to eccentricity while stress on piles 2 and 3 would decrease.

$$\sigma_1 = \frac{R1}{A} = \frac{C}{nA} \pm \frac{C \cdot e \cdot a_1}{I} \qquad (1)$$

$$\uparrow \qquad\qquad \uparrow$$

Stress on a pile if there   Additional stress
is no eccentricity   due to eccentricity

σ_1 = stress on pile 1; A = area of the pile; C = column load
n = number of piles; e = eccentricity

a_1 = the distance between the center of pile 1 and the center of gravity of the pile group, measured parallel to e (eccentricity). (See Figure 8.9.)

I = moment of inertia of the pile group, measured from the axis perpendicular to the eccentricity (going through the center of gravity).

In this case the Y-axis is perpendicular to the eccentricity. Hence all distances should be obtained from the Y-axis.

STEP 2: Compute I (moment of inertia).

Moment of inertia $= A \cdot r^2$ (r = distance to the pile from the axis. In this case the axis perpendicular to the eccentricity passing through the center of gravity should be considered.)

$$I = A \times 2.5^2 + A \times 2.5^2 + A \times 2.5^2 + A \times 2.5^2 = 25\ A$$

(Note: The above distances to the piles are measured from the axis perpendicular to e passing through the center of gravity. The measurement should be taken parallel to e. In this case distance to all piles were taken from the Y-axis measured parallel to eccentricity).

STEP 3: Compute a_1.

a_1 = the distance between the center of pile 1 and the center of gravity of the pile group, measured parallel to e (eccentricity).
$a_1 = 2.5\,\text{ft}$

STEP 4: Apply Eq. (1):

$$\sigma_1 = \frac{R_1}{A} = \frac{C}{nA} \pm \frac{C \cdot e \cdot a_1}{I}$$

$$\frac{R_1}{A} = \frac{10}{4A} + \frac{10 \times 1 \times 2.5}{25A} = \frac{3.5}{A}$$

$$R_1 = 3.5\,\text{tons}$$

(Note: Above the "+" sign was used since pile 1 was on the same side as the column load. The location of the load enhances the load on pile 1. If the 10-ton column load was applied at the center of gravity, then all four piles would be loaded equally.)

By symmetry, load on pile 4 is the same as pile 1. Hence, $R_1 = R_4 = 3.5\,\text{tons}$.

STEP 5: Compute the load on pile 2.

$$\sigma_2 = \frac{R_2}{A} = \frac{C}{nA} \pm \frac{C \cdot e \cdot a_2}{I}$$

$\sigma_2 =$ stress in pile 2; $A =$ area of the pile; $C =$ column load
$n =$ number of piles; $e =$ eccentricity
$a_2 =$ The distance between the center of pile 2 and the center of gravity of the pile group, measured parallel to e (eccentricity)
$I =$ moment of inertia of the pile group, measured around the axis perpendicular to the eccentricity going through the center of gravity

$$\frac{R_2}{A} = \frac{10}{4A} + \frac{10 \times 1 \times 2.5}{25A} = \frac{1.5}{A}$$

In this case $(-)$ sign is used, since the bending moment due to eccentricity tends to reduce the load on pile 2. $R_2 = 1.5$ tons.

By symmetry, the load on pile 3 is also 1.5 tons.

Hence, $R_1 = 3.5$ tons, $R_2 = 1.5$ tons, $R_3 = 1.5$ tons, and $R_4 = 3.5$ tons. (Total equals to 10 tons.)

8.3   Double Eccentricity

Design Example 2

It is possible to have a column load eccentric to both the X- and Y-axis as shown in Figure 8.10. The column load is given as 20 tons. Find the load on each pile.

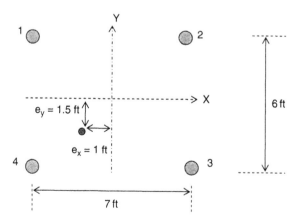

Figure 8.10   Eccentric load location and pile group dimensions

Solution

STEP 1: If the column load is to act on the center of gravity of piles, all four columns will get 5 tons each.

- Assume loads on piles to be R1, R2, R3, and R4, respectively. The area of pile is considered to be A.

$$\sigma_1 = \frac{R_1}{A} = \frac{C}{nA} \pm \frac{C \cdot e_x \cdot a_x}{I_y} \pm \frac{C \cdot e_y \cdot a_y}{I_x} \qquad (2)$$

$$\uparrow \qquad\qquad\qquad \uparrow$$

*Additional stress on pile 1,   Additional stress on pile 1,
due to eccentricity along the   due to eccentricity along the
X-axis   Y-axis*

I_y => In the second term, I_y is used instead of I_x. The moment of inertia should be considered along the axis perpendicular to the eccentricity. If eccentricity is along the X-axis, then the moment of inertia should be considered along the Y-axis.

σ_1 = stress in pile 1; A = area of the pile; C = column load
n = number of piles; e_x = eccentricity parallel to X-axis (e_x = 1 ft)
e_y = eccentricity parallel to Y-axis (e_y = 1.5 ft)
a_x = the distance between the center of pile 1 and the center of gravity of the pile group, measured parallel to e_x (a_x = 3.5 ft)
a_y = the distance between the center of pile 1 and the center of gravity of the pile group, measured parallel to e_y (a_y = 3 ft)

STEP 2: Compute I_x and I_y.

$$I_y = A \cdot 3.5^2 + A \cdot 3.5^2 + A \cdot 3.5^2 + A \cdot 3.5^2 = 49\,A$$
(Distances are measured from the Y-axis.)
$$I_x = A \cdot 3.0^2 + A \cdot 3.0^2 + A \cdot 3.0^2 + A \cdot 3.0^2 = 36\,A$$
(Distances are measured from the X-axis.)
a_x = 3.5 ft   and a_y = 3.0 ft for pile 1.
(a_x is measured parallel to the X-axis, and a_y is measured parallel to the Y-axis.)
e_x = 1 ft   and e_y = 1.5 ft
(These two values are given.)

STEP 3: Apply Eq. 2 to pile 1.

$$\sigma_1 = \frac{R_1}{A} = \frac{C}{nA} \pm \frac{C \cdot e_x \cdot a_x}{I_y} \pm \frac{C \cdot e_y \cdot a_y}{I_x}$$

$$\sigma_1 = \frac{R_1}{A} = \frac{20}{4A} + \frac{20 \cdot (1\,\text{ft}) \cdot (3.5\,\text{ft})}{49\,A} - \frac{20 \cdot (1\,\text{ft}) \cdot (3.0\,\text{ft})}{36\,A}$$

$$\sigma_1 = 3.93/A \text{ tons/sq ft;} \quad R_1 = 3.93 \text{ tons}$$

Eccentricity e_x increases the load on pile 1, while eccentricity e_y decreases the load on pile 1.

STEP 4: Apply the above equation to pile 2.

$$\sigma_2 = \frac{R_2}{A} = \frac{C}{nA} \pm \frac{C \cdot e_x \cdot a_x}{I_y} \pm \frac{C \cdot e_y \cdot a_y}{I_x}$$

$$\sigma_2 = \frac{R_2}{A} = \frac{20}{4A} \pm \frac{20 \cdot (1\,\text{ft})(3.5\,\text{ft})}{49A} - \frac{20 \cdot (1.5\,\text{ft}) \cdot (3.0\,\text{ft})}{36A}$$

Note: Both signs are (−) since both eccentricities are moving away from pile 2. (See Figure 8.10.)

$$\sigma_2 = 1.07/A \text{ tons/sq ft;} \quad R_2 = 1.07 \text{ tons}$$

STEP 5: Apply the above equation to pile 3.

$$\sigma_3 = \frac{R_3}{A} = \frac{C}{nA} \pm \frac{C \cdot e_x \cdot a_x}{I_y} \pm \frac{C \cdot e_y \cdot a_y}{I_x}$$

$$\sigma_3 = \frac{R_3}{A} = \frac{20}{4A} - \frac{20 \cdot (1\,\text{ft}) \cdot (3.5\,\text{ft})}{49A} - \frac{20 \cdot (1.5\,\text{ft}) \cdot (3.0\,\text{ft})}{36A}$$

$$\sigma_3 = 6.07/A \text{ tons/sq ft;} \quad R_3 = 6.07 \text{ tons}$$

For pile 3, eccentricity e_x reduces the load on pile 3, while eccentricity e_y increeas the load on pile 3.

STEP 6: Apply the above equation to pile 4.

$$\sigma_4 = \frac{R_4}{A} = \frac{C}{nA} \pm \frac{C \cdot e_x \cdot a_x}{I_y} \pm \frac{C \cdot e_y \cdot a_y}{I_x}$$

$$\sigma_4 = \frac{R_4}{A} = \frac{20}{4A} \pm \frac{20 \cdot (1\,\text{ft}) \cdot (3.5\,\text{ft})}{49A} - \frac{20 \cdot (1.5\,\text{ft}) \cdot (3.0\,\text{ft})}{36A}$$

$$\sigma_4 = 8.93/A \text{ tons/sq ft;} \quad R_4 = 8.93 \text{ tons}$$

Both eccentricities increase the load on pile 4.
Check whether the sum of four piles adds to 20 tons.
$R_1 = 3.93$; $R_2 = 1.07$; $R_3 = 6.07$; $R_4 = 8.93$

8.4   Pile Groups in Clay Soils

Piles in a group can fail as a group; this type of failure is known as group failure.

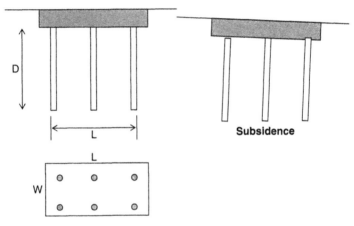

Figure 8.11   Subsidence of a pile group

Group failure as shown above does not occur in sandy soils and is not a design consideration.

Design Methodology for Pile Group Failure

Assume a pile group with dimensions of $L \times W \times D$ as shown above.

$$\text{Skin friction of the group} = 2 \cdot (L + W) \times D \times \alpha \cdot C_u$$

α = skin friction coefficient; C_u = cohesion of the clay soil
End bearing of the group = $N_c \cdot C_u \times (L \times W)$
Pile group capacity = $2 \cdot (L + W) \times D \times \alpha\, C_u + N_c \cdot C_u \times (L \times W)$
N_c = bearing capacity factor = usually taken as 9

Pile Group Capacity Based on Individual Pile Capacity

Pile group capacity based on capacity of individual piles
= Efficiency $\times$ N $\times$ single pile capacity
N = number of piles in the group

Design Example

Find the allowable pile capacity of the pile group shown. Individual piles are of 1-ft diameter. Pile spacing = 4 ft.

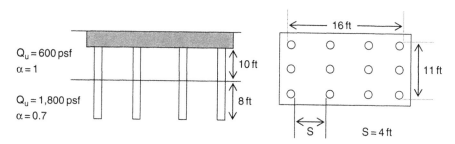

Figure 8.12   Pile group and soil profile

Note: (1) Pile spacing (S) is normally in the range of 3d to 5d. (d = pile diameter)

STEP 1: Find the ultimate pile capacity of an individual pile.

Top layer → $Q_u = 600$ psf; $Q_u/2 = S_u = C = 300$ psf $= 0.15$ tsf;

$\alpha = 1.0$ (α value can be obtained from the Kolk and Van der Valde table.)
Skin friction due to top layer $= \alpha \cdot C \times$ perimeter $= 1.0 \times 0.15 \times \pi \times 1 \times 10 = 4.7$ tons

Bottom layer → $Q_u = 1800$ psf; $Q_u/2 = S_u = C = 900$ psf;
$C = 0.45$ tsf $\alpha = 0.7$

Skin friction due to bottom layer $= \alpha \cdot C \times$ perimeter $= 0.7 \times 0.45 \times \pi \times 1 \times 8 = 7.9$ tons
Total skin friction (for an individual pile) $= 7.9 + 4.7 = 12.6$ tons
Tip bearing resistance $= 9 \cdot C$. area of the pile at the tip $= 9 \times 0.45 \times (\pi \times 1^2)/4 = 3.2$ tons
Tip bearing resistance $= 3.2$ tons

Ultimate pile capacity of an individual pile $= 3.2 + 12.6 = 15.8$ tons

Assume an efficiency factor of 0.8. (the efficiency factor is used because the piles in the group are closely spaced. Hence, piles could stress the soil more than individual piles located far apart.)

Ultimate capacity of the group $= 0.8 \times 12 \times 15.8$ tons $= 151.7$ tons

Note: AASHTO (4.5.6.4) recommends an efficiency factor of 0.7 for pile groups in clay soils with a center-to-center spacing of 3D or more.

STEP 2: Find the ultimate pile capacity of the group as a whole.

- The pile group can fail as a group. This type of failure is known as *group failure.*

The ultimate capacity of the group = Group skin friction

+ Group tip bearing resistance

- The pile group is considered as one big rectangular pile with dimensions of 11ft × 16 ft.

Group skin friction = Group skin friction (top layer)

+ Group skin friction (bottom layer)

Group skin friction (top layer) = $\alpha \cdot C \cdot$ perimeter area of the group

within the top layer

$= 1.0 \times 0.15 \times 2 \times (16 + 11) \times 10$

$= 81$ tons

The perimeter area of the pile group is 2 × (16+11) × 10 (Depth = 10 ft)

Group skin friction (bottom layer) = $\alpha \cdot C \cdot$ perimeter area of the

group within the bottom layer

$= 1.0 \times 0.45 \times 2 \times (16 + 11) \times 8$

$= 194$ tons

For group failure, the α value is taken to be 1.0 since the failure occurs between soil against soil. For soil/pile failure, the α coefficient may or may not be equal to 1.0.

Group tip bearing resistance = $9 \cdot C \times$ (area of the group)

$= 9 \times 0.45 \times (16 \times 11) = 713$ tons

Ultimate group capacity = $81 + 194 + 713 = 988$ tons
Select the lesser of 988 tons and 151.7 tons.
Hence, the ultimate capacity of the group = 151.7 tons
Allowable pile capacity of the group = 151.7/3.0 = 50.6 tons (FOS = 3.0).

Reference

American Association of State Highway and Transportation Officials (AASHTO), *Standard Specifications for Highway Bridges*, 1992.

9

Pile Settlement

9.1 Pile Settlement Measurement

Measurement of settlement in piles is not as straightforward as one would think. When a pile is loaded, two things happen.

1. The pile settles into the soil.
2. The pile material compresses due to the load.

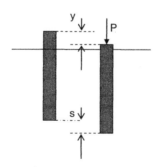

Figure 9.1 Settlement of piles

In Figure 9.1, a pile is loaded with a load P and the settlement at the top (y) is measured. Assume the compression of the pile due to the load to be c.

Due to compression of the pile, y is not equal to s.

$$\text{Settlement } y = s + c \quad \text{(Only y can be measured.)}$$

Why Pile Compression Is Difficult to Compute

Unfortunately, Hook's equation (FL/Ae) = E cannot be used to compute the compression of the pile. This is due to skin friction acting on the pile walls.

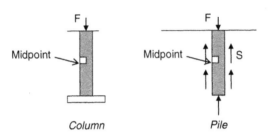

Column   Pile

Figure 9.2 Building column and a pile

1. Figures 9.1 and 9.2 show a column and a pile. At midpoint, a strain gauge is attached to both the pile and the column. (Strain gauge would measure the strain at the point of contact. The stress at that point is calculated using the measured strain value.)

2. The stress at the midpoint of the column = F/A (A = cross-sectional area)

3. The stress at the midpoint of the pile = (F-S)/A
   (A = cross-sectional area, S = total skin friction above midpoint)
   S = Unit skin friction (f) × Pile perimeter × Depth to the midpoint

1. The stress along a column is a constant. The stress along a pile varies with the depth since total skin friction above a given point depends on the depth to that point measured from the top of the pile.

2. Due to skin friction, Hook's equation FL/Ae = E cannot be used directly for piles.

Method to Compute the Settlement and Pile Compression

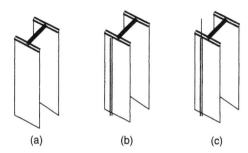

Figure 9.3   Settlement measurement. (a): A regular H-pile is shown. (b): A pipe is welded to the H-pile. (c): A rod is inserted into the pipe. (The above procedure can be done for steel pipe piles, concrete piles, or timber piles.)

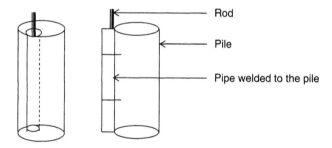

Figure 9.4   Settlement measurement in a pipe pile

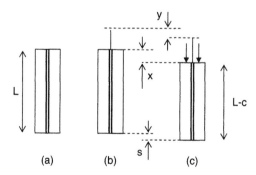

Figure 9.5   Pile settlement diagram

STEP 1: An elevation view of an H-pile with a welded pipe is shown in the left diagram in Figure 9.5.

STEP 2: A rod is inserted into the welded pipe.

STEP 3: The pile is loaded. When the pile is loaded, the pile and the welded pipe compress due to the load. The rod is not loaded. The rod is freely sitting inside the welded pipe. The length of the rod will not change.

STEP 4: y and x can be measured; c and s need to be computed.

STEP 5: $y = s$ The rod is not compressed. Hence, if the rod goes down by s, at the bottom, that same amount will go down at the top of the rod.

STEP 6: Total settlement of the pile at the top (x) = settlement of the pile into the soil (s) + elastic compression of the pile (c)

$x = s + c$ (from the above relationship)

$y = s$ (from step 5)

Hence, $x = y + c$

x and y can be measured. Hence, c can be computed.

9.2   Stiffness of Single Piles

Stiffness of a Spring

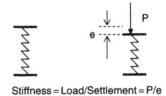

Stiffness = Load/Settlement = P/e

Figure 9.6   Springs

Stiffness of Soil—Pile System

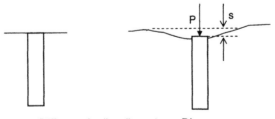

Stiffness of soil—pile system = P/s

Figure 9.7   Stiffness of Soil — Pile System $= P/s$

Soil settles by a distance of s. This settlement occurs for two reasons.

1. Shortening of the pile due to the load

2. Settlement of surrounding soil

In this analysis, settlement due to 1 is ignored. Settlement of the pile due to settlement of soil would be analyzed.

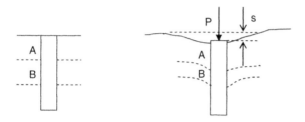

Figure 9.8   Settlement of single pile

Two hypothetical soil layers are selected. When the pile is loaded, the pile stresses the soil immediately next to the pile wall. Hence, the soil immediately next to the wall settles. Soil particles further away from the pile feel less stress; hence, these soil particles settle less. Experiments have shown, that in most cases, soil particles lying at a distance that equals to the length of the pile can be considered to be unstressed.

Assume the shear stress between soil particles and pile wall to be f. (f is the skin friction of the pile.) Skin friction is dependent on the depth in sandy soils.

Shear force acting on a unit length of the pile $= f \times$ pile perimeter $= f \times 2\pi \times r_0$ ($r_0 =$ radius of the pile)

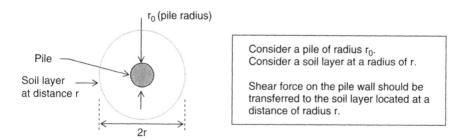

Figure 9.9   Shear zone around the pile

f = shear stress at the pile wall; fs = shear stress of soil at a distance r

$f \times 2\pi \times r_0 = fs \times 2\pi \times r$ $fs = (f \times r_0)/r$

Shear modulus (G) is defined as → G = shear stress/shear strain

(Note: Shear modulus (G) = E/2 (1 + v) (E = Young's modulus; v = Poisson's ratio)

Assume the shear strain at $r = \gamma$.

Hence,

$$G = \frac{\text{Shear stress}}{\text{Shear strain}} = \frac{f_s}{\gamma} = \frac{f \times r_0}{r\gamma}$$

$$\gamma = \frac{f r_0}{G \cdot r}$$

Assume the settlement at a distance r to be s.

Hence, shear strain $= \gamma = ds/dr$

$$ds = \gamma \cdot dr$$

$$S = \int_{r_0}^{r_1} \frac{(f \times r_0)}{Gr} \cdot dr$$

r_1 = influence zone (r_1 varies with the type of soil and pile diameter. Usually, r_1 is taken to be the length of the pile. Shear stress beyond that distance is negligible in most cases)

$$S = \int_{r_0}^{r_1} \frac{(f \times r_0) dr}{G \cdot r} = \frac{(f \times r_0) \ln(r_1/r_0)}{G}$$

9.3 Settlement of Single Piles (Semi-empirical Approach)

The following semi-empirical procedure can be used to compute the settlement of single piles.

The settlement of single piles can be broken down into three distinct parts.

- Settlement due to axial deformation

- Settlement at pile point

- Settlement due to skin friction

Let's look at each component.

9.3.1   Settlement Due to Axial Deformation

As described in this chapter under "Load Distribution of Piles," it is not easy to accurately compute the load distribution along the length of the pile. Skin friction acting on the sidewalls of the pile absorbs a certain percentage of the load. In some cases, skin friction is so paramount that very little load develops at the tip of the pile. Axial compression of the pile is directly linked to the load distribution of the pile.

The following semi-empirical equation is provided to compute the axial compression of piles

$$S_a = (Q_p + a \cdot Q_s)L/AE_p \quad (Vesic, 1977)$$

S_a = settlement due to axial compression of the pile
Q_p = load transferred to the soil at tip level
  $a = 0.5$ for clay soils
  $a = 0.67$ for sandy soils
Q_s = total skin friction load
  L = length of the pile
  A = cross-sectional area of the pile
E_p = Young's modulus of pile
Compare the above equation with axial compression in a free-standing column.
Axial compression of a column $(s) = FL/AE$ (elasticity equation)

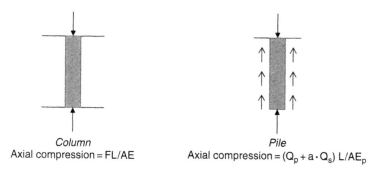

Column
Axial compression $= FL/AE$

Pile
Axial compression $= (Q_p + a \cdot Q_s) L/AE_p$

Figure 9.10   Loading on a building column and a pile

Skin friction acting along pile walls is the main difference between piles and columns.

9.3.2 Settlement at Pile Point

Because of the load transmitted at the pile tip, soil just under the pile tip could settle.

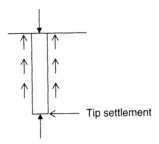

Figure 9.11   Tip settlement

Settlement at pile tip $(S_t) = (C_p \cdot Q_p)/(B \cdot q_0)$ (*Vesic, 1977*)

C_p = empirical coefficient;   B = pile diameter;
q_0 = ultimate end bearing capacity
Q_p = point load transmitted to the soil at pile tip

Table 9.1   Typical values of C_p

Soil Type	Driven Piles	Bored Piles
Dense sand	0.02	0.09
Loose sand	0.04	0.18
Stiff clay	0.02	0.03
Soft clay	0.03	0.06
Dense silt	0.03	0.09
Loose silt	0.05	0.12

Note: Bored piles have higher C_p values compared to driven piles. Higher C_p values would result in higher settlement at the pile point.

9.3.3 Settlement Due to Skin Friction

Skin friction acting along the shaft may stress the surrounding soil. Skin friction acts in an upward direction along the pile. Equal and opposite force acts on the surrounding soil. The force due to pile on surrounding soil would be in a downward direction.

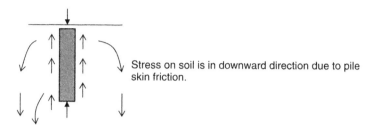

Stress on soil is in downward direction due to pile skin friction.

Figure 9.12   Skin friction development in a pile

When the pile is loaded, the pile moves down slightly. The pile drags the surrounding soil with it. Hence, pile settlement occurs due to skin friction.

$$\text{Settlement due to skin friction } (S_{sf}) = (C_s \cdot Q_s)/(D \cdot q_0)$$

C_s = empirical coefficient = $(0.93 + 0.16 \times D/B)\, C_p$
C_p values are obtained from the Table 9.1.
Q_s = total skin friction load
D = length of the pile; B = diameter of the pile

References

NAVFAC DM-7, *Foundations and Earth Structures*, U.S. Department of the Navy, 1982.
Vesic, A.S., *Design of Pile Foundations*, National Cooperative Highway Research Program 42, Transportation Research Board, 1977.

9.4   Pile Settlement Comparison (End Bearing vs. Floating)

• Most engineers are reluctant to recommend floating piles due to high settlement compared to end bearing piles.

• This chapter compares floating piles and end bearing piles.

Factors that Affect Settlement

The following parameters affect the settlement of piles:

L/D ratio (L = length of the pile and D = diameter of the pile)
Ep/Es ratio (Ep and Es = Young's modulus of pile and soil, respectively)

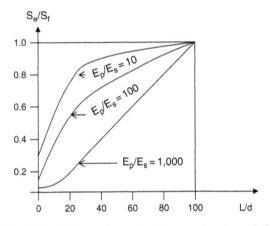

Figure 9.13   Pile settlement graph. (*Source*: Poulos, H.G, 1989)

S_e = Settlement of an end bearing pile; S_f = Settlement of a floating pile

End bearing pile (settlement = S_e)   *Floating pile (settlement = S_f)*

Figure 9.14   End bearing and floating piles

Figure 9.14 shows the relationship between the S_e/S_f ratio and L/d ratio for a given load.

Design Example

A 30-ft-long pile with a diameter of 1 ft is placed on bedrock. The E_p/E_s ratio was computed to be 100. If the pile is designed as a floating pile, find the increase of settlement.

Solution

$L/d = 30$; $E_p/E_s = 100$
From the graph; $S_e/S_f = 0.62$; $S_f = S_e/0.62 = 1.61\ S_e$

A floating pile would have a 60% higher settlement than the end bearing pile.
The following observations could be made from the graph in Figure 9.14.

- When the L/d ratio increases, the settlement difference between two pile types becomes negligible. For an instance, when L/d is 100, S_e/S_f ratio reaches 1. This means that it is not profitable to extend long piles all the way to the bedrock.

- When the Ep/Es ratio increases, the Se/Sf ratio decreases.

Assume a pile with an L/d ratio of 20 and an $\frac{E_p}{E_s}$ ratio of 100. This pile would have a $\frac{S_e}{S_f}$ ratio of 0.5. Assume that the pile material is changed and the new $\frac{E_p}{E_s}$ ratio is 1,000. Then the $\frac{S_e}{S_f}$ ratio would change to 0.2. This means that it is not very profitable to extend a pile with a low Ep/Es ratio to bedrock.

$L/d = 20$ and $E_p/E_s = 100$; then $S_e/S_f = 0.5$ or $S_f = 2 \times S_e$
$L/d = 20$ and $E_p/E_s = 1000$; then $S_e/S_f = 0.2$ or $S_f = 5 \times S_e$

9.5   Critical Depth for Settlement

- Pile settlement can be reduced by increasing the length of the pile. It is been reported that increasing the length beyond the critical depth will not cause a reduction of settlement.

Critical depth for settlement is given by the following equation:

$$L_c/d = \{(\pi \cdot E_p \cdot A_p)/(E_s \cdot d^2)\}^{1/2} \quad (Hull, 1987)$$

L_c = critical depth for settlement; E_p = Young's modulus of pile material
E_s = Young's modulus of soil; A_p = area of the pile;
d = pile diameter

- Critical depth for settlement has no relationship to critical depth for end bearing and skin friction.

References

Butterfield, R., and Ghosh, N., "The Response of Single Piles in Clay to Axial Load," Proc. 9th Int. Natl. Conf., Soil Mechanics and Foundation Eng., Tokyo, 1, 451–457, 1998.

Hull, T.S., "The Static Behavior of Laterally Loaded Piles," Ph.D. Thesis, University of Sydney, Australia, 1987.

Poulos, H.G., "Pile Behavior, Theory and Application," Geotechnique, 366–413, 1989.

9.6   Pile Group Settlement in Sandy Soils

The following equation could be adopted to compute pile group settlement in sandy soils.

$$S_g = S(B/D)^{1/2} \quad (Vesic, 1977)$$

S_g = settlement of the pile group
S = settlement of a single pile
B = smallest dimension of the pile group
D = diameter of a single pile

Design Example

A (3×3) pile group (with 1-ft-diameter piles) is loaded with 270 tons. It is assumed that the load is uniformly distributed among all piles. Settlement of a single pile due to a load of 30 tons is calculated to be 1 in. Estimate the pile group settlement.

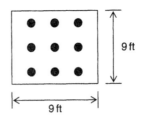

Figure 9.15 (9×9) Pile group

$S_g = S (B/D)^{1/2}$
$B = 9$ ft; $D = 1$ ft; $S = 1$ in.
$S_g = 1$ in. $\times (9/1)^{1/2} = 3$ in.

When a single pile is loaded to 30 tons, the settlement was 1 in. When the same pile is loaded to 30 tons inside a group, the settlement becomes 3 in. (Total load on the group is 270 tons. There are nine piles in the group. Hence, each pile is loaded to 30 tons.)

Larger settlement in a pile group can be attributed to the shadowing effect.

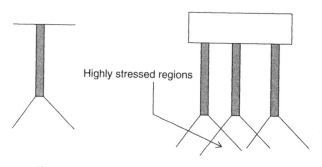

Figure 9.16 Stressed region in a pile group

References

NAVFAC DM-7, *Foundations and Earth Structures*, U.S. Department of the Navy, 1982.

Vesic, A.S., "Design of Pile Foundations," National Cooperative Highway Research Program 42, Transportation Research Board, 1977.

9.7 Long-Term Pile Group Settlement in Clay Soils

Pile groups may undergo consolidation settlement in clay soils. Consolidation settlement of pile groups is computed using the following simplified assumptions.

- Assume the pile group to be a solid foundation with a depth of two-thirds the length of piles.

- Effective stress at the midpoint of the clay layer is used to compute settlement.

- Consolidation equation:

    Consolidation settlement $(S) = C_c/(1 + e_0) \cdot H \cdot \log_{10} \cdot (p_0' + \delta_p)/p_0'$

C_c = compression index
e_0 = initial void ratio of the clay layer
H = thickness of the clay layer
p_0' = initial effective stress at midpoint of the clay layer
δ_p = increase of effective stress due to pile load

Design Example

A pile group consists of nine piles each, with a length of 30 ft and a diameter of 1 ft. The pile group is 10 ft × 10 ft and is loaded with 160 tons. Density of soil is 120 pcf.

The following clay parameters are given:

C_c = compression index = 0.30
e_0 = initial void ratio of the clay layer = 1.10
Depth to bedrock from the bottom of the piles = 20 ft
Groundwater is at a depth of 8 ft below the surface.
Find the long-term settlement due to consolidation.

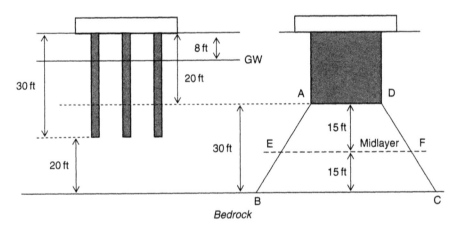

Figure 9.17   Load distribution underneath a pile group

Solution

STEP 1: Simplify the pile group to a solid footing with a depth of 20 ft (two-thirds of the length of piles).

STEP 2: Find the effective stress at midlayer of the clay stratum. The midlayer occurs 15 ft below the end of the assumed solid footing.

$$p_0' = (8 \times 120) + 27 \times (120 - 62.4) = 2,515.2 \text{ psf}$$

$(8 \times 120) =$ effective stress component above the groundwater level $27 \times (120 - 62.4) =$ effective stress component from the groundwater level to the midlayer

STEP 3: Find the effective stress increase due to the pile group.

Assume a 2V to 1H stress distribution.
Pile group dimensions $= 10 \text{ ft} \times 10 \text{ ft}$
Length EF $= 10 + 2 \times (15/2) = 25 \text{ ft}$ (see Figure 9.17)
Pile group load $= 160$ tons
Stress at bottom of assumed solid footing $= 160/(10 \times 10) = 1.6 \text{ tsf}$
Stress at midlayer $160/(25 \times 25) = 0.256 \text{ tsf} = 512 \text{ psf}$
$\delta_p = 512 \text{ psf}$

STEP 4: Consolidation settlement of the pile group $(S) = C_c /(1 + e_0) \cdot H \cdot \log_{10} \cdot (p_0' + \delta p)/p_0'$

$C_c = 0.30$
$e_0 = 1.10$
$H =$ thickness of the compressible portion of the clay layer $= 30 \text{ ft}$
$S = 0.30/(1 + 1.10) 0.30 \cdot \log_{10} \cdot (2,515 + 512)/2,515 = 0.345 \text{ ft} = 4.13 \text{ in.}$

Note: The consolidation settlement can be reduced by conducting any of the following:

a) Increase the length of piles. (When the length of piles is increased, the H value in the above equation will reduce. H is the thickness of the compressible portion of the clay layer.)

b) Increase the length and width of the pile group. (When the length and width of the pile group are increased, stress increment (δ_p) due to pile group load would decrease.)

9.8 Long-Term Pile Group Settlement in Clay Soils—Janbu Method

The method explained in Chapter 9.7 needs C_c value and e_0 value. Laboratory tests are needed to obtain these parameters. This chapter presents a simplified procedure to deduce long-term settlement in pile groups. The parameters in the Janbu method are easily obtained and compared to the compression index method described in the previous chapter.

- The Janbu method can be used to compute the long-term elastic settlement in both sandy soils and clayey soils.
- Janbu assumed that the soil settlement is related to two nondimensional parameters.
- Stress exponent—j (nondimensional parameter)
- Modulus number—m (nondimensional parameter)
- The stress exponent (j) and modulus number (m) are unique to individual soils.

Janbu Equation for Clay Soils

$$\text{Settlement} = L \times 1/m \ \ln(\sigma_1'/\sigma_0')$$

Since stress exponent, $j = 0$ for pure clay soils, it does not appear in the above equation.

The modulus number is obtained from Table 9.2. Note that j is not equal to zero for silty clays and clayey silts. (See Table 9.2.)

L = thickness of the clay layer; $\ln$ = natural log
m = the modulus number (dimensionless)
σ_1' = new effective stress after the pile load
σ_0' = effective stress prior to the pile load (original effective stress)

Table 9.2 Janbu soil settlement parameters

Soil Type	Modulus Number (m)	Stress Exponent (j)
Till (very dense to dense)	1,000–300	1
Gravel	400–40	0.5
Sand (dense)	400–250	0.5
Sand (medium dense)	250–150	0.5
Sand (loose)	150–100	0.5
Silt (dense)	200–80	0.5
Silt (medium dense)	80–60	0.5
Silt (loose)	60–40	0.5

Table 9.2—Cont'd

Soil Type	Modulus Number (m)	Stress Exponent (j)
Silty Clay (or Clayey Silt)		
Stiff silty clay	60–20	0.5
Medium silty clay	20–10	0.5
Soft silty clay	10–5	0.5
Soft marine clay	20–5	0
Soft organic clay	20–5	0
Peat	5–1	0

Settlement Calculation Methodology

The pile group is represented with an equivalent footing located at the neutral plane. The neutral plane is considered to be located at 2/3H from the pile cap (H = pile length).

The settlement of the pile group is computed using the Janbu equation.

Design Example

Estimate the settlement of the pile group shown. (H = 30 ft and D = 10 ft) (Density of organic clay = 110 pcf; Column load = 100 tons; Pile cap size = 5 ft × 5ft)

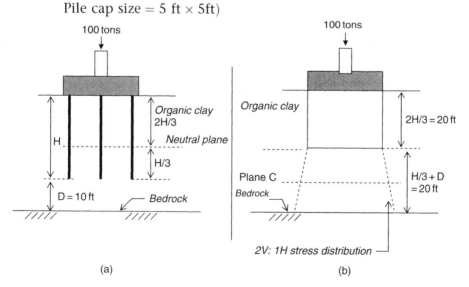

Figure 9.18 Pile group settlement example

STEP 1: Consider a footing located at the neutral plane.

1. The stress at the neutral plane $= 100/(5 \times 5)$ tsf $= 4$ tsf

2. The thickness of the compressible clay layer $= H/3 + D = (30/3 + 10)$ ft $= 20$ ft

STEP 2: Apply the Janbu equation.

$$\text{Janbu Equation} \rightarrow \text{Settlement} = L \times 1/m \; \ln(\sigma_1'/\sigma_0')$$

$L =$ thickness of the clay layer $= 20$ ft
$m =$ Table PG –1 gives a range of 20 to 5 for soft organic clay. Use $m = 10$.
$\sigma_0' =$ initial effective stress at midpoint of the compressible clay layer. The midpoint of the compressible clay layer occurs at plane C, at a depth of 30 ft below the ground surface.
$\sigma_0' = 30 \times 110 = 330$ psf
$\sigma_1' = 330$ psf + Stress increase due to column load at the midpoint. (1)
Column load at neutral plane $= 100$ tons. (The width and length of the equivalent footing at the neutral plane is 5 ft $\times$ 5 ft.)
Length at plane C $= 5$ ft $+ 10/2 + 10/2 = 15$ ft (assuming a 2 Vertical to 1 Horizontal stress distribution)
Hence, the area at plane C $= 15$ ft $\times 15$ ft $= 225$ sq ft
Stress increase due to column load at point C $= 100/225 = 0.444$ tsf $= 889$ psf
$\sigma_1' = 330$ psf + Stress increase due to column load at the midpoint. [From above Eq. (1)]
$\sigma_1' = 330 + 889 = 1{,}219$ psf

$$\text{Janbu equation} \rightarrow \text{Settlement} = L \times 1/m \; \ln(\sigma_1'/\sigma_0')$$

Settlement $= 20 \times 1/10 \ln (1219/330) = 2.61$ ft

This is a very high settlement; hence, increase the area of the pile cap or embedment depth of piles.

9.9 Pile Group Settlement in Sandy Soils

- Janbu proposed the following semi-empirical equation for sandy soils. Sandy soils reach the maximum settlement within a short period of time.

$$\text{Settlement} = L \times (2/m) \times [(\sigma_1'/\sigma_r)^{1/2} - (\sigma_0'/\sigma_r)^{1/2}] \text{ tsf}$$

Note that the stress exponent (j) for sandy soils is 0.5 and that it is already integrated into the above equation. The above equation is unit sensitive. All parameters should be in tons and feet for English units or meters and kPa for metric units.

L = thickness of the compressible sand layer
m = the modulus number (dimensionless)
σ_1' = new effective stress after the pile load
σ_0' = effective stress prior to the pile load (original effective stress)
σ_r = reference stress (100 kPa for metric units and or 1 tsf for English units)

Janbu Procedure for Sandy Soils

• Assume the neutral axis to be at a depth two-thirds the length of the piles.

• The pile group is simplified to a solid footing ending at a depth of two-thirds the length of the piles.

• Compute the original stress (stress due to overburden prior to the construction of piles).

• Compute the stress increase due to pile group.

• Use the Janbu equation to find the settlement.

Design Example

Estimate the settlement of the pile group shown. (H = 30 ft and D = 10 ft)
   (Column load = 100 tons; Pile cap size = 5 ft × 5 ft, Density of medium dense sand = 100 pcf).

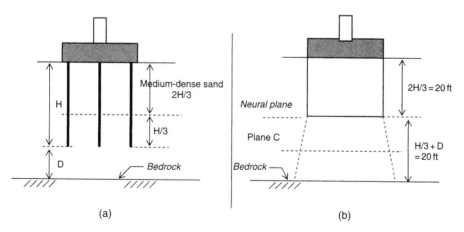

(a)   (b)

Figure 9.19   Neutral plane concept

STEP 1: Simplify the pile group to a solid footing ending at two-thirds the depth of the pile group.

The stress at the neutral plane $= 100/(5 \times 5)$ tsf $= 4$ tsf

1. The thickness of the compressible sand layer $= H/3 + D = (30/3 + 10)$ ft $= 20$ ft

STEP 2: Apply the Janbu equation.

Janbu equation $\rightarrow$ Settlement $= L \times (2/m) \times [(\sigma_1'/\sigma_r)^{1/2} - (\sigma_0'/\sigma_r)^{1/2}]$ tsf (1)

$L =$ thickness of the sand layer $= 20$ ft
$m =$ Table 9.2 gives a range of 250 to 150 for medium-dense sand. Use $m = 200$.
$\sigma_0' =$ initial effective stress at midpoint of the compressible sand layer

The midpoint of the compressible sand layer occurs at plane C, at a depth of 30 ft below the ground surface.

$\sigma_0' = 30 \times 100 = 300$ psf $= 0.15$ tsf. (All units should be converted to tsf.)
$\sigma_1' = 0.15$ tsf + Stress increase due to column load at the midpoint. (2)

Column load at neutral plane $= 100$ tons. (the width and length of the equivalent footing at the neutral plane is 5 ft $\times$ 5 ft.)

Length of footing at plane C $= 5$ ft $+ 10/2 + 10/2 = 15$ ft

Hence, the area at plane C $= 15$ ft $\times 15$ ft $= 225$ sq ft

Stress increase due to column load at point C $= 100/225 = 0.444$ tsf

$\sigma_1' = 0.15$ tsf + Stress increase due to column load at the midpoint. [From Eq. (2)]

$\sigma_1' = 0.15 + 0.444 = 0.594$ tsf

When computing the stress increase due to the column load at midpoint, stress distribution of 2V:1H is used.

Settlement $= L \times (2/m) \times [(\sigma_1'/\sigma_r)^{1/2} - (\sigma_0'/\sigma_r)^{1/2}]$ tsf

$\sigma_r =$ reference stress $= 1$ tsf

Settlement $= 20 \times 2/200 \times [(0.594)^{1/2} - (0.15)^{1/2}]$ tsf $= 0.077$ ft $= 0.92$ in.

Reference

Winterkorn, H.F., and Fang, H.H., Foundation Engineering Handbook, Van Nostrand Reinhold Co., New York, 1975.

9.10   Pile Group Settlement vs. Single Pile Settlement

Researchers in the past did not see any relationship between settlement of single piles and settlement of pile groups. The following comment was made by Karl Terzaghi regarding this issue.

> *Both theoretical considerations and experience leave no doubt that there is no relation whatever between the settlement of an individual pile at a given load and that of a large group of piles having the same load per pile.*
>
> *James Forrest Lecture, Karl Terzaghi (1939)*

Thanks to numerous analytical methods and high-powered computers, researchers have been able to develop relationships between single-pile settlement and pile group settlement.

Factors that Affect Pile Group Settlement

L/d ratio (L = pile length; d = pile diameter)
s/d ratio (s = spacing between piles)
E_p/E_s ratio (E_p = Young's modulus of pile; E_s = Young's modulus of soil)

Interestingly, the geometry of the group does not have much of an influence on the settlement. A (2 × 8) group and a (4 × 4) group would act in almost the same manner.

Group Settlement Ratio (Rs)

Group Settlement Ratio (Rs) = (Settlement of Group/Settlement of Single Pile)

R_s can be approximated as follows:
$R_s = n^{0.5}$ (n = number of piles in the group)

Design Example

A single pile has a settlement of 1.5 in. when it is loaded to 100 tons. A (4 × 4) pile group (of similar piles) was loaded to 1,600 tons. Find the settlement of the group.

Solution

Settlement of the group = Rs × Settlement of a single pile
Rs = $n^{0.5} = 16^{0.5} = 4$
Settlement of the group = 4 × 1.5 in. = 6 in.

The pile group settles more than a single pile when the load per pile is the same as the single pile. This happens due to the increase of soil stress within the group.

Consolidation Settlement

1. Settlement due to consolidation is more important for pile groups, than for single piles. Hence, it is recommended to compute the settlement due to consolidation as well as the elastic settlement.

References

Terzaghi, K., "Soil Mechanics—A New Chapter in Engineering Science," *J. Inst. of Civil Engineers,* 12, 106–141, 1939.
Poulos, H.G., "Pile Behavior, Theory and Application," *Geotechnique,* 366–413, 1989.

Pile Group Design (Capacity and Settlement)—Example

Design Example

Find the allowable pile capacity and the settlement of the pile group shown. Individual piles are of 1-ft diameter. Pile group efficiency $= 0.75$.

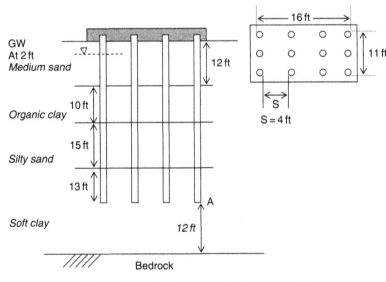

Figure 9.20   Pile group design example

The following table provides soil parameters for the soil strata in Figure 9.20.

	Medium Sand	Organic Clay	Silty Sand	Soft Clay
Thickness (ft)	12	10	15	25
Density (pcf)	120	115	118	110
K (Earth pressure coefficient)	1		1.2	
δ (Friction angle between pile and soil)	25		30	
φ (Soil friction angle)	30		35	
α		1		0.65
(C) Cohesion (psf)		500 psf		120 psf
Nc				9
M (Janbu parameter)	200	10	80	15
J (Janbu parameter)	0.5	0	0.5	0.5

STEP 1: Compute the skin friction of an individual pile:

Skin friction in sands → S (skin friction) $= K \times \sigma \times \tan \delta$
$\times$ Perimeter of the pile

Skin friction in clays → S (skin friction) $= \alpha \times c$
$\times$ Perimeter of the pile

- Skin friction in first layer (medium sand) $= K \times \sigma \times \tan \delta \times [(\pi \times 1) \times 12]$ at mid point of the pile in the given layer is used.

- $= 120 \times 2 + (120 - 62.4) \times 4 = 470.4$ psf   (Groundwater is at 2 ft.)

  Skin friction in first layer (medium sand) $= 1 \times 470.4 \times \tan 25 \times [(\pi \times 1) \times 12] = 7{,}345$ lbs

- Skin friction in second layer (organic clay) $= \alpha \times C \times [(\pi \times 1) \times 10] = 1 \times 500 \times [(\pi \times 1) \times 10] = 15{,}708$ lbs

- Skin friction in third layer (silty sand) $= K \times \sigma \times \tan\delta \times [(\pi \times 1) \times 15]$ σ at midpoint of the pile in the given layer is used

  $\sigma = 120 \times 2 + (120 - 62.4) \times 10 + (115 - 62.4) \times 10 + (118 - 62.4) \times 7.5$
  $= 1{,}759$ psf

  Skin friction in third layer (silty sand) $= 1.2 \times 1759 \times \tan 30 \times [(\pi \times 1) \times 15] = 50{,}682$ lbs

- Skin friction in fourth layer (soft clay) $= \alpha \times C \times [(\pi \times 1) \times 13] = 0.65 \times 120 \times [(\pi \times 1) \times 13] = 3{,}185$ lbs

Total skin friction $= 7345 + 15708 + 50682 + 3185 = 76{,}920$ lbs

STEP 2: Compute the ultimate end bearing capacity (acting as individual piles).

Ultimate end bearing capacity in clay $= N_c \times C \times$ Pile tip area $(N_c = 9)$
Ultimate end bearing capacity in soft clay $= 9 \times 120 \times \pi \times$ Diameter$^2/4$
  $= 848.2$ lbs/per pile
Total ultimate bearing capacity per pile $= 76{,}920 + 848.2 = 77{,}768$ lbs
Total ultimate bearing capacity of the group (assume a group efficiency
  of 0.8) $= 12 \times 77{,}768 \times 0.8 = 746{,}573$ lbs $= 373$ tons

STEP 3: Compute the pile group capacity (failure as a group)

- Skin friction of a pile group in clay $= C \times$ Perimeter of the pile group
  Since the failure is between soil and soil, α coefficient is essentially 1.

- Skin friction of a pile group in sand $= K \times \sigma \times \tan \phi \times$ Perimeter of the
  pile group Notice in the case of a pile group that ϕ is used instead of
  δ, since failure is between soil and soil.

- Skin friction in first layer (medium sand) $= K \times \sigma \times \tan \phi \times [16 + 16 + 11 + 11] \times 12$

  σ at midpoint of the pile in the given layer is used $= 120 \times 2 + (120 - 62.4) \times 4 = 470.4$ psf

- Skin friction in first layer (medium sand) $= 1 \times 470.4 \times \tan 30 \times 54 \times 12 = 155{,}313$ lbs

- Skin friction in second layer (organic clay) $= C \times [16 + 16 + 11 + 11] \times 10 = 500 \times 54 \times 10 = 270{,}000$ lbs

- Skin friction in third layer (silty sand) $= K \times \sigma \times \tan \phi \times [16 + 16 + 11 + 11] \times 15$ σ at midpoint of the pile in the given layer is used.

  $\sigma = 120 \times 2 + (120 - 62.4) \times 10 + (115 - 62.4) \times 10 + (118 - 62.4) \times 7.5 = 1{,}759$ psf

  Skin friction in third layer (silty sand) $= 1.2 \times 1759 \times \tan 30 \times 54 \times 15 = 871{,}160$ lbs

- Skin friction in fourth layer (medium stiff clay) $= C \times [16 + 16 + 11 + 11] \times 13 = 120 \times 54 \times 13 = 84{,}240$ lbs

Total skin friction $= 155,313 + 270,000 + 871,160 + 84,240 =$ 1,380,713 lbs $= 690$ tons

- End bearing capacity of the pile group in clay $= N_c \times C \times$ Area $(N_c = 9)$ $= 9 \times 120 \times (16 \times 11) = 190,080$ lbs $= 95$ tons

- Total ultimate capacity $= 690 + 95 = 785$ tons

In this case, the pile group capacity acting as a group is greater than the collective capacity of individual piles. Hence, the lower value of 373 tons is used with a factor of safety of 2.5.
Allowable load on the pile group $= 150$ tons

STEP 4: Settlement Computation

- The Janbu tangent modulus procedure is used to compute the pile group settlement. Consider an equivalent shallow foundation placed at a neutral plane (at a depth of 2H/3 from the surface). (H $=$ length of piles)

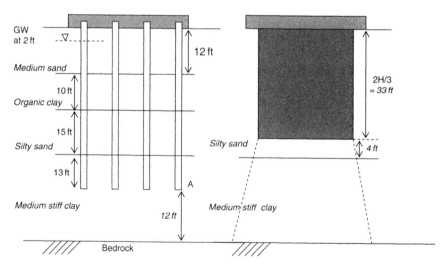

Figure 9.21   Pile group settlement example

- Compute the settlement of the 4-ft-thick silty sand layer.

Janbu Equation for Sandy Soils

$$\text{Settlements} = L \times 2/m[(\sigma_1')^{1/2} - (\sigma_1')^{1/2}] \text{ tsf}$$

L = thickness of the compressible sand layer;

m = the modulus number (dimensionless)

σ_1' = new effective stress after the pile load

σ_0' = effective stress prior to the pile load (original effective stress)

L = thickness of the sand layer = 4 ft

m = Table 9.2 gives a range of 250 to 150 for medium dense sand. Use $m = 200$.

σ_0' = initial effective stress at midpoint of the compressible sand layer. The midpoint of the compressible sand layer occurs 2 ft below the hypothetical footing.

$\sigma_0' = 2 \times 118 = 236$ psf $= 0.118$ tsf. (All units should be converted to tsf.)

$\sigma_1' = 0.118$ tsf + Stress increase due to column load at the midpoint.

Column load at neutral plane = 150 tons. (Width and length of the equivalent footing at the neutral plane is 11 ft × 16 ft.)

Length of base at midpoint of the silty sand layer = 16 ft + 2/2 + 2/2 = 18 ft

Width of base at midpoint of the silty sand layer = 11 ft + 2/2 + 2/2 = 13 ft

Hence, area at midplane of silty sand layer = 18 ft × 13 ft = 234 sq ft

Stress increase due to column load at midpoint = 150/234 = 0.64 tsf

$\sigma_1' = 0.118$ tsf + Stress increase due to column load at the midpoint.

$\sigma_1' = 0.118 + 0.64 = 0.758$ tsf

γ When computing the stress increase due to the column load at midpoint, stress distribution of 2V:1H is used.

Janbu equation → Settlement $= L \times 2/m\left[(\sigma_1')^{1/2} - (\sigma_0')^{1/2}\right]$ tsf

Settlement $= 2 \times 2/200\left[(0.758)^{1/2} - (0.118)^{1/2}\right]$ tsf $= 0.011$ ft $= 0.92$ in.

- Settlement in medium-stiff clay

Janbu Equation for Clay Soils

$$\text{Settlements} = L \times 1/m \; [\ln (\sigma_1'/\sigma_0')]$$

$L =$ thickness of the clay layer; $\ln =$ natural log
$m =$ the modulus number (dimensionless)
$\sigma_1' =$ new effective stress after the pile load
$\sigma_0' =$ effective stress prior to the pile load (original effective stress)

Consider a footing located at neutral plane.

Thickness of the compressible clay layer $= 25$ ft
$m =$ Table 9.2 gives a range of 20 to 10 for medium silty clay. Use $m = 15$.
$\sigma_0' =$ initial effective stress at midpoint of the compressible clay layer.
  Midpoint of the medium silty clay layer occurs at a depth of 49.5 ft
  from the surface.
$\sigma_0' = 2 \times 120 + 10 \times (120 - 62.4) + 10 \times (115 - 62.4) + 15 \times (118 - 62.4) +$
  $12.5 \times (110 - 62.4) = 2{,}771$ psf $= 1.385$ tsf
$\sigma_1' = \sigma_0' +$ Stress increase due to column load at the midpoint
Area at the midpoint of the medium-stiff clay layer $= (16 + 16.5/$
$2 + 16.5/2) \times (11 + 16.5/2 + 16.5/2) = 893.75$ sq ft
$\sigma_1' = \sigma_0' + 150/893.75 = 1.385 + 0.1678 = 1.552$ tsf
Settlement in medium-stiff clay $= 25 \times 1/15 \; [\ln (1.552/1.385)] =$
  0.189 ft $= 2.28$ in.

Settlement due to both medium-stiff clay and silty sand $= 2.28 +$
$0.92 = 3.2$ in.

This settlement is excessive. Piles need to be driven deeper, or the
number of piles needs to be increased.

10

Pile Design in Rock

10.1 Rock Coring and Logging

The following information has been found to be valuable for most projects.

Rock Joints

A rock joint is basically a fracture in the rock mass. Most rocks consist of joints. Joints can occur in the rock mass for many reasons.

- Earthquakes—Major earthquakes can shatter the bedrock and create joints.
- Plate tectonic movements—Continents move relative to each other. When they collide, the bedrock folds and joints are created.
- Volcanic eruptions.
- Generation of excessive heat in the rock.

Depending on the location of the bedrock, the number of joints in a core run can vary. Some core runs contain few joints, whereas other core runs may contain dozens of joints.

Joint Set

When a group of joints are parallel to each other, that group of joints is called a *joint set*.

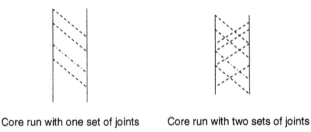

Core run with one set of joints   Core run with two sets of joints

Figure 10.1 Rock cores with joints

Joint Filler Materials

Some joints are filled with matter. Typical joint filler materials are

- *Sands*: Sands occur in joints where there is high-energy flow (or high velocity).
- *Silts*: Silts indicate that the flow is less energetic.
- *Clays*: Clay inside joints indicates stagnant water in joints.

Joint filler material information would be very useful in interpreting Packer test data. It is known that filler material could clog up joints and reduce the flow rate with time.

Rock Joint Types

Most joints can be divided into two types.

- Extensional joints (joints due to tensile failure)
- Shear joints

Slickensided (very smooth) joint surfaces indicate shearing. Smooth planar joints also most probably could be shear joints.

Core Loss Information

Experts have stated that core loss information is more important than the rock core information. Core loss location may not be obvious in most cases. Coring rate, color of return water, and arrangement of core in the box can be used to identify the location of the core loss.

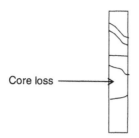

Figure 10.2   Rock core loss

Core loss occurs in weak rock or in highly weathered rock.

Fractured Zones

A fracture log can provide fractured zones. Fracture logs of each boring can be compared to check for joints.

Drill Water Return Information

The engineer should be able to assess the quantity of returning drill water. Drill water return can vary from 90 to 0%. This information can be very valuable in determining weak rock strata. Typically, drill water return is high in sound rock.

Water Color

The color of returning drill water can be used to identify the rock type.

• Rock Joint Parameters

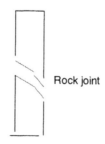

Figure 10.3   Rock joint

Joint Roughness

Joint surface can be rough or smooth. Smooth joints can be less stable than rough joints.

Joint Alteration

Alterations to the joint, such as color and filling of materials.

Joint Filler Material

Some joints can be filled with sand, while other joints can get filled with clay. Smooth joints filled with sand may provide additional friction. However, rough joints filled with clay may reduce the friction in the joint.

Joint Stains

Joint stains should be noted. Stains can be due to groundwater and various other chemicals.

- Joint Types

  Extensional joints: Joints that form due to tension or pulling apart.
  Shear joints: Joints that occur due to shearing.

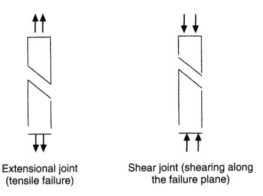

Extensional joint
(tensile failure)

Shear joint (shearing along
the failure plane)

Figure 10.4 Extensional joints and shear joints

Dip Angle and Strike

Each set of joints would have a dip angle and a strike.

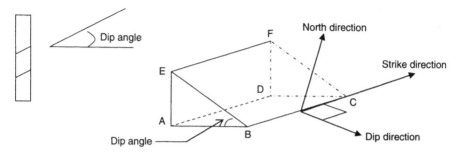

Figure 10.5   Dip and strike

Joint Plane

EBCF is the joint plane in Figure 10.5.

Dip Angle

The dip angle is the angle between the joint plane (EBCF) and the horizontal plane (ABCD). This angle is easily measured in the field.

Strike

The strike line is the horizontal line BC as shown in the figure.

Strike Direction

The right-hand rule is used to obtain the strike direction.

- Open your right hand and face the palm down. Mentally lay the palm on the joint plane.

- Point four fingers (except the thumb) along the downward direction of the slope.

- The direction of the thumb indicates the strike direction.

- The clockwise angle between the strike direction and the North direction is called the strike angle. The ABCD plane and North direction are in the same horizontal plane.

Dip Direction

The dip direction is the direction of the downward slope and is different from the dip angle. Strike and dip directions are perpendicular to each other.

Notation: Dip angle and strike angle are written as shown below.

35/100

The above first value indicates the dip angle (not the dip direction). The second value indicates the strike direction measured clockwise from the North. Dip angle is always written with two digits, while strike direction is always written with three digits. A joint plane with a dip angle of 40° and a 70° strike angle is written as 40°/070°.

10.2 Oriented Rock Coring

Oriented Coring Procedure

STEP 1: A knife is installed in the core barrel to create a mark in the rock core. This mark is known as the scribe mark.

STEP 2: The knife is installed in such a manner that a line drawn through the scribe mark and the center of the core would point toward the North direction (line AB). The driller would use a magnetic compass, prior to coring and locate the North direction. Then he would be able to locate the knife.

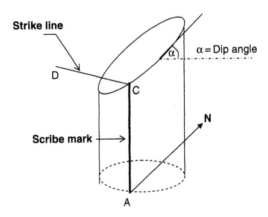

Figure 10.6 Oriented coring

STEP 3: Draw a horizontal line going through point C, at the joint (line CD). Both lines (line AN and line CD) are in horizontal planes.

STEP 4: Measure the clockwise angle between line AN (which is the North direction) and line CD. This angle is the strike angle of the joint.

Oriented Coring Procedure (Summary)

- The driller finds the North direction using a compass. Then the driller places the knife along the North direction. The line connecting the knife point (point A) and the center of the core will be the North-South line (line AN).

- Rock coring is conducted. A scribe mark will be created along the rock core. (The rock core does not rotate during coring.)

- After the rock core is removed from the hole, a horizontal line is drawn along the plane of the joint (line CD). The clockwise angle between line AN and line CD is the strike angle.

10.3   Oriented Core Data

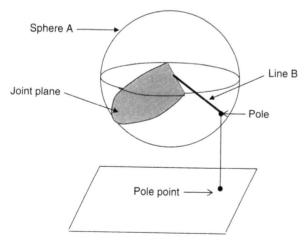

Figure 10.7   Oriented coring geometry

- Oriented coring would produce a *dip angle* and a *strike angle* for each joint.

- A "joint" can be represented by one point known as the pole.

Concept of Pole

- STEP 1: The joint plane is drawn across the sphere A.

- STEP 2: A perpendicular line is drawn (line B) to the joint plane.

- STEP 3: The point where line B intersects the sphere is known as the pole.

- STEP 4: A vertical line is dropped from the pole to obtain the pole point.

Theoretically, there are two poles for any given joint (one in the lower hemisphere and the other one in the upper hemisphere). The pole in the lower hemisphere is always selected.

As you can see, it is not easy to obtain the pole point for a given joint. For that purpose, one needs a sphere and has to go through lot of trouble. Charts are available to obtain the pole point for any given joint.

10.4   Rock Mass Classification

Who is the better athlete?

Athlete A: Long Jump: 23 ft, runs 100 m in 11 sec, High jump 6.5 ft
Athlete B: Long Jump: 26 ft, runs 100 m in 13 sec, High jump 6.4 ft

Athlete A is very weak in long jump but very strong in 100 meters. Athlete B is very good in long jump but not very good in 100 meters. Athlete A has a slight edge in high jump.

It is not easy to determine which athlete is better. For this reason, the Olympic committee came up with a marking system for the decathlon. The best athlete is selected based on the marking system.

A similar situation exists in rock types. Let's look at an example.

Example

A geotechnical engineer has the choice to construct a tunnel in either rock type A or rock type B.

Rock Type A → Average RQD = 60%, joints are smooth, joints are filled with clay
Rock Type B → Average RQD = 50%, joints are rough, joints are filled with sand

Rock type A has a higher RQD value. On the other hand, Rock type B has rougher joints. Smooth joints in rock type A are not favorable for geotechnical engineering work. Joints in rock type A are filled with clay, while joints in rock type B are filled with sands. Based on the above information, it is not easy to select a candidate, since both rock types have good properties and bad properties. Early engineers recognized the need of a classification system to determine the better rock type. Unfortunately, more than one classification system exist. These systems are named *Rock Mass Classification Systems.*

Popular Rock Mass Classification Systems are:

- Terzaghi Rock Mass Classification System (rarely used)

- Rock Structure Rating (RSR) method by Wickham (1972)

- Rock Mass Rating system (RMR) by Bieniawski (1976)

- Rock Tunneling Quality Index by Barton et al. (1972). (Better known as the Q system)

Recently, the Q system has gained popularity over other systems.

10.5   Q System

$$Q = \frac{RQD}{J_n} \times \frac{J_r}{J_a} \times \frac{J_w}{SRF}$$

Q = Rock Quality Index; RQD = Rock Quality Designation
J_n = joint set number; J_r = joint roughness number; J_a = joint alteration number
J_w = joint water reduction factor; SRF = stress reduction factor

Rock Quality Designation (RQD)

To obtain RQD, select all the rock pieces longer than 4 in. in the rock core. Measure the total length of all the individual rock pieces greater than 4 in. This length is given as a percentage of the total length of the core.

Total length of the core = 60 in.
Total length of all the pieces longer than 4 in. = 20 in.
RQD = 20/60 = 0.333 = 33.3%

RQD (0%–25%) = Very poor
RQD (25%–50%) = Poor
RQD (50%–75%) = Fair
RQD (75%–90%) = Good
RQD (90%–100%) = Excellent

Joint Set Number (J_n)

Find the number of joint sets. When a group of joints have the same dip angle and a strike angle, that group is known as a joint set. In some cases many joint sets exist. Assume there are eight joints in the rock core with the following dip angles: 32, 67, 35, 65, 28, 64, 62, 30, 31. It is clear that there are at least two joint sets. One joint set has a dip angle approximately at 30°, while the other joint set has a dip angle of approximately at 65°.

The Q system allocates the following numbers:

Zero Joints → $J_n = 1.0$
One Joint Set → $J_n = 2$
Two Joint Sets → $J_n = 4$
Three Joint Sets → $J_n = 9$
Four Joint Sets → $J_n = 15$

The higher J_n number indicates a weaker rock for construction.

Joint Roughness Number (J_r)

When subjected to stress, smoother joints may slip and failure can occur before rougher joints. For this reason, joint roughness plays a part in rock stability.

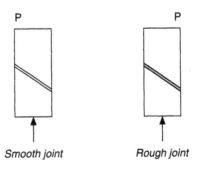

Smooth joint   Rough joint

Figure 10.8 Smooth and rough joints

Note: Slippage occurs along a smoother joint at a lower load (P).

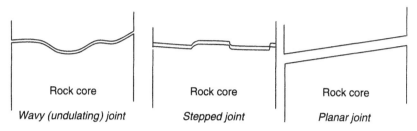

Rock core   Rock core   Rock core

Wavy (undulating) joint   *Stepped joint*   *Planar joint*

Figure 10.9   Rock joint types

How to Obtain the Joint Roughness Number?

STEP 1: There are three types of joint surface profiles.

• Wavy (undulating) joint surface profiles.

• Stepped joint surface profiles.

• Planar joint surface profiles.

No joint surface is either 100% planar, stepped, or wavy. Select the type that best describes the joint surface (see Figure 10.9).

STEP 2: Feel the joint surface and categorize into one of the following types.

• *Rough*—If the surface feels rough

• *Smooth*—If the surface feels smooth

• *Slickensided*—Slickensided surfaces are very smooth and slick. They occur when there is a shear movement along the joint. In some cases, the surface could be polished. One would notice polished (shining) patches. These polished patches indicate shear movement along the joint surface. The name "slickensided" was given to indicate slick surfaces.

STEP 3: Use the following table to obtain J_r.

Table 10.1   Joint roughness coefficient (J_r)

Joint Profile	Joint Roughness	J_r
Stepped	Rough	4
	Smooth	3
	Slickensided	2
Undulating	Rough	3
	Smooth	2
	Slickensided	1.5
Planar	Rough	1.5
	Smooth	1.0
	Slickensided	0.5

Source: Hoek, Kaiser and Bawden (1975).

- Rock joints with stepped profiles provide the best resistance against shearing. From the above table smooth joint with a stepped profile would be better than a rough joint with a planar profile.

Joint Alteration Number (J_a)

Joints get altered with time. They get altered due to material filling inside them. In some cases filler material could cement the joint tightly. In other cases, filler material could introduce a slippery surface creating a much more unstable joint surface.

Table 10.2   Joint alteration number (J_a)

Description of Filler Material	J_a
(A) Tightly healed with a nonsoftening impermeable filling seen in joints (*quartz or epidote*)	0.75
(B) Unaltered joint walls. No filler material seen (surface stains only)	1.0
(C) Slightly altered joint walls. Nonsoftening mineral coatings are formed; sandy particles, clay, or disintegrated rock are seen in the joint.	2.0
(D) Silty or sandy clay coatings, small fraction of clay in the joint	3.0
(E) Low *friction* clay in the joint (Kaolinite, talc, and chlorite are low friction clays.)	4.0

Source: Hoek, Kaiser and Bawden (1975).

It is easy to notice a tightly healed joint. In this case use $J_a = 0.75$. If the joint has not undergone any alteration other than surface stains, use $J_a = 1.0$. If there are sandy particles in the joint, then use $J_a = 2.0$. If there is clay in the joint, then use $J_a = 3.0$. If clay in the joint can be considered low friction, use $J_a = 4.0$. For this purpose, clay type existing in the joint needs to be identified.

Joint Water Reduction Factor (J_w)

J_w cannot be obtained from boring data. A tunnel in the rock needs to be constructed to obtain it. Usually data from previous tunnels constructed in the same formation is used to obtain J_w. Another option is to construct a pilot tunnel ahead of the real tunnel.

Table 10.3   Joint water reduction factor (J_w)

Description	Approximate Water Pressure (Kg/cm^2)	J_w
(A) Excavation (or the tunnel) is dry. No or slight water flow into the tunnel.	Less than 1.0	1.0
(B) Water flows into the tunnel at a medium rate (water pressure 1.0–2.5 kgf/cm²). Joint fillings get washed out occasionally due to water flow.	−2.5	0.66
(C) Large inflow of water into the tunnel or excavation. The rock is competent and joints are unfilled (water pressure 1.0–2.5 kgf/cm²).	2.5–10.0	0.5
(D) Large inflow of water into the tunnel or excavation. The joint filler material gets washed away (water pressure 2.5–10 kgf/cm²).	Greater than 10	0.33
(E) Exceptionally high inflow of water into the tunnel or excavation (water pressure > 10 kgf/cm²).	Greater than 10	0.1 to 0.2

Source: Hoek, Kaiser and Bawden (1975).

Stress Reduction Factor (SRF)

SRF cannot be obtained from boring data. A tunnel in the rock needs to be constructed to obtain SRF as in the case of J_w. Usually, data from previous tunnels constructed in the same formation is used to obtain SRF. Another option is to construct a pilot tunnel ahead of the real tunnel.

All rock formations have weak zones, that is, a region in the rock formation which has a low RQD value. Weak zones may have weathered rock or different rock type.

Table 10.4 Stress reduction factor (SRF)

More than one weak zone occur in the tunnel. In this case use SRF = 10.0.
Single weak zone of rock with clay or chemically disintegrated rock. (Excavation depth < 150 ft). Use SRF = 5.0.
Single weak zone of rock with clay or chemically disintegrated rock. (Excavation depth > 150 ft). Use SRF = 2.5.
More than one weak zone of rock without clay or chemically disintegrated rock. Use SRF = 7.5.
Single weak zone of rock without clay or chemically disintegrated rock. (Excavation depth < 150 ft). Use SRF = 5.0.
Single weak zone of rock without clay or chemically disintegrated rock. (Excavation depth > 150 ft). Use SRF = 2.5.
Loose open joints observed. Use SRF = 5.0.

Design Example 1

The average RQD of a rock formation was found to be 60%. Two sets of joints have been identified. Most joint surfaces are undulated (wavy) and rough. Most joints are filled with silts and sands. It has been reported that medium inflow of water has occurred during the construction of past tunnels. During earlier constructions, a single weak zone containing clay was observed at a depth of 100 ft. Find the Q value.

STEP 1:

$$Q = \frac{RQD}{J_n} \times \frac{J_r}{J_a} \times \frac{J_w}{SRF}$$

$RQD = 60\%$
Since there are two sets of joints $J_n = 4$.
$J_r = 3$ (from Table 10.1, for undulating, rough joints).
Since most joints are filled with silts and sands, $J_a = 2$ (from Table 10.2).
$J_w = 0.66$ (from Table 10.3).
$SRF = 5$ (from Table 10.4).
Hence, $Q = (60/4) \times (3/2) \times (0.66/5) = 2.97$.

References

Hoek, E., Kaiser, P.K., and Bawden, W.F., "Support of Underground Excavations in Hard Rock," A.A Balkeema Publishers, 1995.

10.6   Caisson Design in Rock

Caissons under Compression

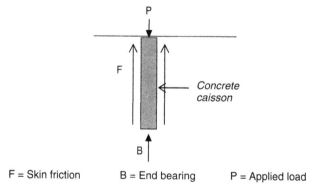

Figure 10.10   Loading on a caisson

Note: Most of the load is taken by skin friction in rock. In most cases, end bearing is less than 5% of the total load. In other words, 95% or more load will be carried by skin friction.

Design Example 2

Design a concrete caisson with a W–section (steel) at the core to carry a load of 1,000 tons. Assume the skin friction to be 150 psi and end bearing to be 200 psi.

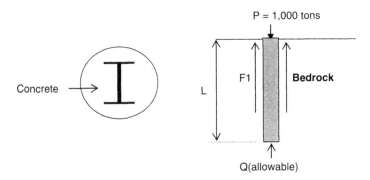

Figure 10.11   Concrete caisson with an I section

The following parameters are given:

Ultimate Steel Compressive Strength = 36,000 psi
Ultimate Concrete Compressive Strength = 3,000 psi

Simplified Design Procedure

A Simplified design procedure is explained first. In this procedure, the composite nature of the section is ignored.

STEP 1: Structural Design of the Caisson

Assume a diameter of 30 in. for the concrete caisson. Since the E (elastic modulus) of steel is much higher than concrete, a major portion of the load is taken by steel. Assume 90% of the load is carried by steel.

Load carried by steel = 0.9 × 1,000 tons = 900 tons
Allowable Steel Compressive Strength = 0.5 × 36,000 = 18,000 psi
Factor of safety of 0.5 is used.

$$\text{Steel area required} = \frac{900 \times 2000}{18,000} = 100 \,\text{sq in}$$

Check the manual of steel construction for an appropriate W section. Use W14 × 342. This section has an area of 101 sq in. The dimensions of this section are given in Figure 10.12.

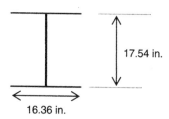

16.36 in.

Figure 10.12   I-section

STEP 2: Check whether this section fits inside a 30-in. hole.

Distance along a diagonal (Pythagoras theorem) $= (16.36^2 + 17.54^2)^{1/2} =$ 23.98 in.
This value is smaller than 30 in. Hence, the section can easily fit inside a 30-in. hole.

STEP 3: Compute the load carried by concrete.

- Concrete area
  = Area of the hole – Area of steel
  = 706.8 – 101 = 605.8

- Allowable concrete compressive strength
  = 0.25 × Ultimate compressive strength
  = 0.25 × 3000 = 750 psi
  Factor of safety of 0.25 is used. Engineers should refer to local Building Codes for the relevant factor of safety values.

- Load carried by concrete
  = Concrete Area × 750 psi
  = 605.8 × 750 lbs = 227.6 tons

- Load carried by steel
  = 900 tons (computed earlier)

- Total capacity of the caisson
  = 900 + 227.6 = 1127.6 tons > 1,000 tons

(Note: The designer can start with a smaller steel section to optimize the above value.)

STEP 4: Compute the required length (L) of the caisson.

- The skin friction is developed along the perimeter of the caisson.
- Total perimeter of the caisson
  $= \pi \times$ (Diameter) $\times$ (Length) $= \pi \times D \times L$
- Total skin friction
  $= \pi \times 30 \times L \times$ Unit skin friction of rock (in this case 150 psi)
- Total skin friction $= \pi \times 30 \times L \times 150$ lbs (L should be in inches)
- Design the caisson so that 95% of the load is carried by skin friction.
  $0.95 \times 1,000$ tons $= 950$ tons
- Hence, the total load carried through skin friction $= 950$ tons

Total skin friction $= \pi \times 30 \times L \times 150 = 950$ tons $= 950 \times 2,000$ lbs
(L) Length of the caisson required $= 134$ in. $= 11.1$ ft

Design Example 3

The following parameters are given.

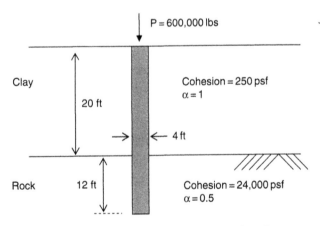

Figure 10.13   Caisson in clay and rock

Caisson diameter $= 4$ ft
Compressive strength of steel $= 36,000$ psi
$E_r/E_c = 0.5$ ($E_r =$ elastic modulus of rock; $E_c =$ elastic modulus of concrete)
Cohesion of the bedrock $= 24,000$ psf

Adhesion coefficient (α) for rock $= 0.5$
Adhesion coefficient (α) for clay $= 1.0$

Solution

STEP 1: Compute the ultimate end bearing capacity:

q_u = Ultimate end bearing strength of the bedrock $= N_c \times$ Cohesion;
   $N_c = 9$
Hence, $q_u = 9 \times$ Cohesion $= 216,000\,psf$
Ultimate end bearing capacity (Q_u) $=$ area $\times q_u$
$Q_u = (\pi \times 4^2/4) \times 216,000 = 1,357$ tons
Allowable end bearing capacity ($Q_{allowable}$) $= 1,357/3 = 452$ tons

STEP 2: Compute the ultimate skin friction:

Ultimate unit skin friction per unit area within the rock mass (f) $= \alpha \cdot C$
   $= \alpha \cdot 24,000$
Adhesion coefficient (α) $= 0.5$
Hence, Ultimate unit skin friction (f) $= 0.5 \times 24,000 = 12,000\,psf$
Assuming a FOS of 3.0, the allowable unit skin friction per unit area
   within the rock mass [$f_{allowable}$] $= 12,000/3.0 = 4,000\,psf = 2\,tsf$.

STEP 3: Find the skin friction within the soil layer.

The skin friction generated within the soil layer can be calculated as in
   a pile.
Soil skin friction (f_{soil}) $= \alpha \cdot C$; $\alpha =$ adhesion factor
(f_{soil}) $= 1.0 \times 250 = 250\,psf$
Skin friction mobilized along the pile shaft within the clay layer $=$
   (f_{soil}) $\times$ Perimeter $= 250 \times (\pi \times d) \times 20 = 62,800\,lbs$
Allowable skin friction $= 62,800/3.0 = 20,900\,lbs = 10$ tons
   (A factor of safety of 3.0 is assumed.)
Load transferred to the rock (F) $= (P - 20,900)\,lbs = 600,000 - 20,900 =$
   $579,100\,lbs$

Note: It is assumed that allowable skin friction within soil is fully mobilized.

STEP 4: Load transferred to the rock

The load transferred to the rock is divided between the total skin
friction (F_{skin}) and the end bearing at the bottom of the caisson [Q_{base}].

(f_{skin}) = unit skin friction mobilized within the rock mass
(F_{skin}) = total skin friction = f_{skin} × perimeter area within the rock mass
(q_{base}) = end bearing stress mobilized at the base of the caisson
(Q_{base}) = q_{base} × area of the caisson at the base

STEP 5: Find the end bearing (Q_{base}) and skin friction (F_{skin}) within the rock mass

The end bearing ratio (n) is defined as the ratio between the end bearing load of the rock mass (Q_{base}) and the total resistive force mobilized $(Q_{base} + F_{skin})$ within the rock mass.
End bearing ratio (n) = $Q_{base}/[Q_{base} + F_{skin}]$; n is obtained from Table 10.5.
Q_{base} = end bearing load generated at the base; F_{skin} = Total skin friction generated within the rock mass.

Table 10.5   End bearing ratio (n)

$E_r/E_c = 0.5$		$E_r/E_c = 1.0$		$E_r/E_c = 2.0$		$E_r/E_c = 4.0$	
L/a	n	L/a	n	L/a	n	L/a	n
1	0.5	1	0.48	1	0.45	1	0.44
2	0.28	2	0.23	2	0.20	2	0.16
3	0.17	3	0.14	3	0.12	3	0.08
4	0.12	4	0.08	4	0.06	4	0.03

Source: Osterberg and Gill (1973).

L = length of the caisson within the rock mass
a = radius of the caisson = 2 ft

$$\frac{E_r}{E_c} = \frac{\text{Elastic modulus of rock}}{\text{Elastic modulus of concrete}} = 0.5 \,(\text{given})$$

Total load transferred to the rock mass = 579,100 lbs = $Q_{base} + F_{skin}$
(see step 3)
Assume a length (L) of 8 ft (since a = 2 ft; L/a = 4)
From Table 10.5, for L/a of 4 and E_r/E_c of 0.5, end bearing ratio (n) = 0.12
n = 0.12 = $Q_{base}/(Q_{base} + F_{skin})$
0.12 = $Q_{base}/579,100$
Hence Q_{base} = 579,100 × 0.12 = 69,492 lbs = 35 tons

$Q_{allowable} = 452$ tons (see step 1)

$Q_{allowable}$ is greater than the end bearing load (Q_{base}) generated at the base.

$F_{skin} = $ load transferred to the rock – end bearing load

$F_{skin} = 579,100 - 69,492 = 509,608\,lbs = 255$ tons

F_{skin} should be less than $F_{allowable}$.

$F_{allowable} = f_{allowable} \times$ perimeter of the caisson within the rock mass

$f_{allowable} = 2\,tsf$ (see step 2)

Since a length (L) of 8 ft was assumed within the rock mass,

$F_{allowable} = 2\,tsf \times (\pi \times 4) \times 8 = 201$ tons.

Skin friction generated $= 255$ tons (see above)

$F_{allowable}$ is less than the skin friction generated. Hence, increase the pile diameter or length of the pile.

Reference

Osterberg, J.O., and Gill, S.A., "Load Transfer Mechanism for Piers Socketed in Hard Soils or Rock," *Proceedings of the Canadian Rock Mechanics Symposium*, 235–261, Montreal, 1973.

PART 3
Design Strategies

11

Lateral Loading Analysis

11.1 Winkler Modulus for Piles

Springs are often used to model the soil–pile interaction. These springs are known as Winkler springs after Winkler who was the first to use springs to model pile behavior.

Modeling of Skin Friction Using Winkler Springs

The skin friction of a pile can be represented using a series of springs.

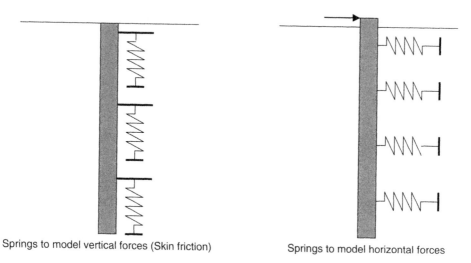

Springs to model vertical forces (Skin friction)   Springs to model horizontal forces

Figure 11.1 Winkler spring model

Spring Constant (k)

- The spring constant (k) is defined as $k = f/w$ (f = force or pressure; w = displacement).

Soil Spring Constant (Coefficient of Subgrade Reaction)

- Coefficient of Subgrade Reaction = Pressure/displacement; (Units – lbs/cu ft)

Methods to Find Coefficient of Subgrade Reaction

All techniques used to find the subgrade modulus can be divided into three categories.

- Experimental methods
- Numerical methods
- Simple theoretical models

Vertical Spring Constant (Vertical Modulus of Subgrade Reaction—k)

$$k/G_s = 1.3 \, (E_p/E_s)^{-1/40} \, [1 + 7 \, (L/D)^{-0.6}] \quad \text{(Mylonakis, 2001)}$$

G_s = soil shear modulus; E_p and E_s = pile and soil Young's modulus;
L = pile length
D = pile diameter

(Note: The above equation should be used only for vertical springs.)

Reference

Mylonakis, G., "Winkler Modulus for Axially Loaded Piles," *Geotechnique*, 455–460, 2001.

11.2 Lateral Loading Analysis—Simple Procedure

- Lateral loads are exerted on piles due to wind, soil, and water. In such cases, lateral pile capacity needs to be designed to accommodate the loading.

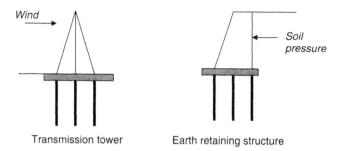

Figure 11.2   Lateral loads

- The deformation of piles due to lateral loading is normally limited to the upper part of the pile. Lateral pile deflection, 8 to 10 diameters below the ground level, is negligible in most cases.

- Piles that can carry heavy vertical loads may be very weak under lateral loads.

11.2.1   Design Methodology of Laterally Loaded Piles

- It is assumed that the pile is being held by springs as shown in Figure 11.3. The spring constant or the coefficient of subgrade reaction varies with the depth. In most cases, the coefficient of subgrade reaction increases with depth.

- Simplified analysis of lateral loads on piles can be conducted by assuming the coefficient of subgrade reaction to be a constant with depth. For most cases, the error induced by this assumption is not significant.

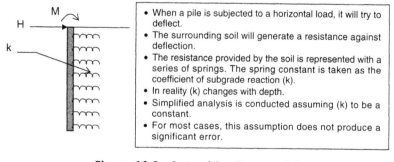

Figure 11.3   Lateral loading model

- The equation for lateral load analysis is as follows:

$$u = (2)^{1/2}(H/k) \times (l_c/4)^{-1} + (M/k)(l_c/4)^{-2} \quad \textit{(Matlock and Reese, 1960)}$$

u = lateral deflection

H = applied lateral load on the pile (normally due to wind or earth pressure)

k = coefficient of subgrade reaction (assumed to be a constant with depth)

M = moment induced due to lateral forces (when the lateral load is acting at a height above the ground level, then moment induced also should be taken into consideration).

l_c = critical pile length (Below this length, the pile is acting as an infinitely long pile.)

l_c is obtained using the following equation:

$l_c = 4 \, [(EI)_p/k]^{1/4}$

$(EI)_p$ = Young's modulus and moment of inertia of the pile. In the case of wind loading, the moment of inertia should be taken against the axis, which has the minimum moment of inertia, since wind load could act from any direction.

In the case of soil pressure and water pressure, the direction of the lateral load does not change. In these situations, the moment of inertia should be taken against the axis of bending.

A similar equation is obtained for the rotational angle (θ) at the top of the pile.

$$\theta = (H/k) \times (l_c/4)^{-2} + (2)^{1/2}(M/k)(l_c/4)^{-3}$$

The derivation of the above equations is provided by Matlock and Reese (1960).

Reference

Matlock, H., and Reese, L.C., "Generalized Solution for Laterally Loaded Piles," *ASCE J. of Soil Mechanics and Foundation Eng.* 86, SM 5 (63–91), 1960.

12

Load Distribution Inside Piles

12.1 Introduction

- Prior to analyzing the load distribution inside a pile, it is important to look at the load distribution inside a building column.

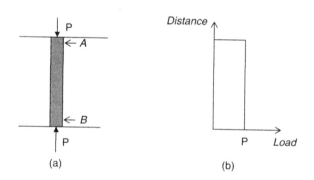

Figure 12.1 Load distribution inside a building column

- Strain gauges can be attached to points A and B, and stress inside the column can be deduced. Multiplication of the stress by the cross-sectional area would provide the internal load in the column.

- Consider point A, immediately below load P. At this location, the column will be stressed to compensate the load from above. Hence, load inside the column will be equal to P.

- This load will be transferred all the way down to the bottom. Load at point B will be equal to P as well. Upward reaction at the footing level also will be equal to P.

Load Distribution inside a Pile Prior to Loading (Sandy Soils)

First, we need to look at the load distribution inside a pile immediately after driving the pile, prior to applying the load.

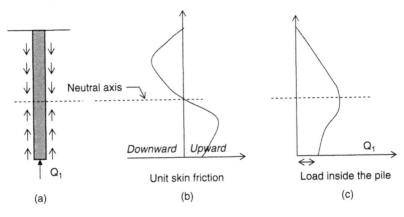

Figure 12.2   Neutral axis

- Figure 12.2a shows a pile immediately after driving, prior to applying the load. Interestingly, there is an end bearing load (Q1) even before any load is applied. This is due to the weight of the pile and downward skin friction. In this example, the weight of the pile is ignored for simplicity.

Elasticity in Soil

- When a pile is driven, the soil around the pile is stressed. Because of elastic forces, surrounding soil tries to push the pile upward.

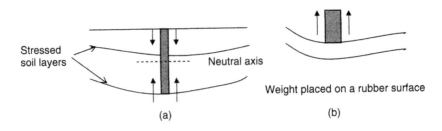

Figure 12.3   Soil stress due to piles

- After pile driving is completed, the bottom soil layers try to push the pile out of the ground due to elastic forces. This action of the bottom soil layers is countered by the action of the top soil layers.

- Think of a weight placed on a rubber mattress as shown above right in Figure 12.3b. The rubber mattress is exerting an upward force on the weight due to elasticity. Similarly, the bottom soil layers are exerting an upward force on the pile.

- Elastic forces acting upward are resisted by the skin friction of top layers. If elastic forces are too large to be countered by skin friction, then the pile will pop out of the ground. This does not usually happen.

Neutral Axis

- The neutral axis is defined as the layer of soil that stays neutral or that does not exert a force on the pile. Reversal of force direction occurs at the neutral axis.

Residual Stresses

- Through elasticity of soil, the pile is stressed just after it has been driven prior to loading. These stresses existing in the pile just after driving, prior to loading, are known as *residual stresses*. In the past, researchers ignored residual stresses inside piles prior to loading. Today it is well accepted that residual stresses could be significant in some soils.

- The difference between the upward and downward load will be transferred to the soil below the pile in the form of end bearing load. (In this case it is taken to be "Q1.")

- Figure 12.2c shows the loads generated inside the pile. The internal pile load at the very top of the pile is zero, since there is no external load sitting on the pile.

- When one moves downward along the pile, the pile starts to get stressed owing to downward skin frictional load. Hence, the internal pile load increases as shown.

- The internal pile load keeps increasing until the neutral axis. At the neutral axis, the internal pile load reaches its maximum. Below the neutral axis, the direction of the skin friction changes. Hence, the internal pile load starts to decrease.

- Bottom of the pile: In sandy soils there is an end bearing load at the bottom of the pile.

- Relationship between applied load (in this case zero) and the end bearing load:
  End bearing load = Applied load + Downward skin friction + Weight of the pile

$$Q_1 = 0 + \text{Downward skin friction} - \text{Upward force due to elasticity of soil}$$

Note: The weight of the pile is ignored.

Load Distribution Inside a Pile (Small Load Applied) (Sandy Soils)

Assume that a small load has been applied to the pile. Load distribution diagrams are as shown in Figure 12.4.

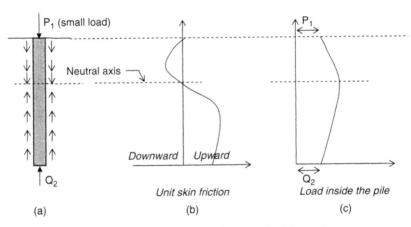

Figure 12.4 Load distribution inside a pile

- When a small load is applied to the pile, the neutral axis moves up. This occurs since the pile has a higher tendency to move downward due to the applied load (P_1). The end bearing load (Q_2) will be higher than the previous case ($Q_2 > Q_1$).

- End bearing load = Applied load + Downward skin friction + Weight of the pile − Upward skin friction

$$Q_2 = P_1 + \text{Downward skin friction} - \text{Upward forces due to elasticity}$$

Note: The weight of the pile is ignored.

- When a load is applied to the pile, the neutral axis moves up. Hence, downward skin friction is reduced.

- When the applied load is increased, the neutral axis moves upward and eventually disappears. At that point there will be no downward skin friction. The total length of the pile will try to resist the applied load.

Load Distribution Inside a Pile (Large Load Applied) (Sandy Soils)

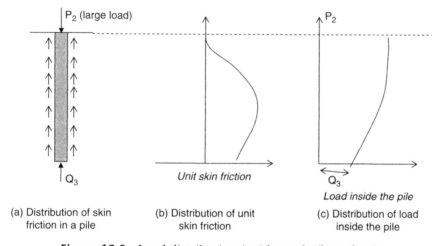

(a) Distribution of skin friction in a pile

(b) Distribution of unit skin friction

(c) Distribution of load inside the pile

Figure 12.5   Load distribution inside a pile (large load)

- When the applied load is increased, the neutral axis moves up. This occurs since the pile has a higher tendency to move downward due to the applied load (P_2). End bearing load (Q_3) will be higher in this case than the previous end bearing load (Q_2).

- End bearing load = Applied load + Weight of the pile − Upward skin friction

$$Q_3 = P_2 + \text{Weight of the pile} - \text{Upward skin friction}$$

(Note: No downward skin friction.)

Residual Stresses in Clay Soils

Residual stress in sandy soils was discussed earlier. Elastic forces similar to sandy soils act on a pile driven in clay soils. In addition to elastic forces, hydrostatic forces are also a factor in clay soils. When a pile is driven, the soil underneath the pile gets pushed down and to the side. This soil movement can create stress in the surrounding soil (see Figure 12.6).

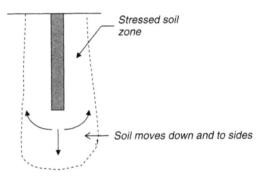

Figure 12.6   Stressed soil around a pile

- Pore pressure in the surrounding soil increases due to induced stress.

- In sandy soils, pore pressure dissipates within minutes, whereas in clay soils it may take months or years.

- When pore pressure dissipates, soil tends to settle. Settling soil around the pile creates a downward force on the pile known as the *residual compression*. Elastic forces act in an upward direction on the pile just after driving, while hydrostatic forces act downward.

12.2   Computation of the Loading Inside a Pile

- Unit skin friction is proportional to the effective stress.

- Hence, unit skin friction at a depth of Z can be represented by kZ. (k is a constant.)

- Total skin friction load at depth Z would be $kZ^2/2$.

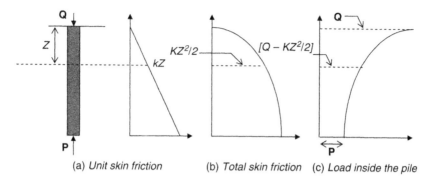

(a) Unit skin friction   (b) Total skin friction   (c) Load inside the pile

Figure 12.7   Skin friction and load distribution

- Figure 12.7a shows the unit skin friction at depth Z (or the skin friction per sq ft at depth Z). The second figure shows the total skin friction load at depth Z. Total skin friction load is equal to the area inside the triangle.

- *Total skin friction at depth Z = $KZ^2/2$*

- Figure 12.7c shows the loading generated inside the pile. Q is the total load applied from the top. At depth Z, soil skin friction has absorbed a load of $kZ^2/2$. Hence, only a load of $[Q - kZ^2/2]$ would be transferred below depth Z.

- *Loading inside the pile at depth Z = $Q - kZ^2/2$*

- The bottom of the pile shows an end bearing of P transferred to the soil underneath, from the pile.

  - The NYC Building Code recommends that 75% of the load be taken by end bearing when the pile tip is located in any type of rock (including soft rock).

13

Neutral Plane Concept

13.1 Introduction

- During the pile-driving process, the soil around the pile is stressed.

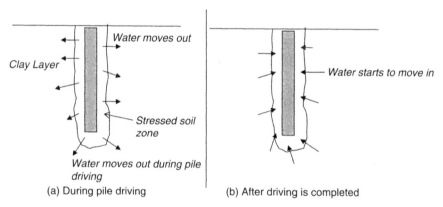

Figure 13.1 Water movement around piles

- Due to the high stress generated in surrounding soil during pile driving, water is dissipated from the soil as shown in Figure 13.1a.

- After pile driving is completed, water starts to come back to the previously stressed soil region around the pile (Figure 13.1b).

- Due to migration of water, soil around the pile consolidates and settles.

- Settlement of soil surrounding the pile is very small, yet large enough to exert a downward force on the pile. This downward force is known as residual compression.

- Due to the downward force exerted on the pile, the pile starts to move downward.

- Eventually the pile comes to an equilibrium and stop.

- At equilibrium, *upper* layers of soil exert a downward force on the pile, while *lower* layers of soil exert an upward force on the pile.

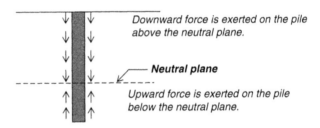

Figure 13.2   Negative skin friction and neutral axis

- The direction of the force reverses at neutral plane.

Soil and Pile Movement (above the neutral plane)

- Both soil and the pile move downward above the neutral plane. But the *soil* moves slightly more in downward direction than the *pile*. Hence, relative to the pile, soil moves downward.

- This downward movement of soil relative to the pile can exert a downward drag on the pile.

Soil and Pile Movement (below the neutral plane)

- Both soil and the pile move downward below the neutral plane as in the previous case.

- But this time, the pile moves slightly more in a downward direction relative to the soil. Hence, relative to the pile, soil moves upward.

- This upward movement of soil relative to the pile exerts an upward force on the pile.

Soil and Pile Movement (at the neutral plane)

- Both soil and the pile move downward at the neutral plane as in previous cases.

- But at the neutral plane both soil and pile move downward by the same margin.

- Hence, relative movement between soil and pile is zero.

- At the neutral plane, no force is exerted on the pile.

Location of the Neutral Plane

- The exact location of the neutral plane cannot be estimated without elaborate techniques involving complicated mathematics.

- For floating piles, the neutral plane is taken at two-thirds of the pile length.

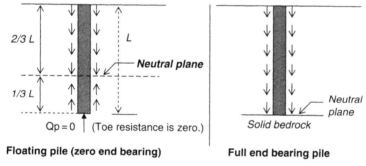

Figure 13.3   Floating piles and end bearing piles

- In the case of full end bearing piles (piles located on solid bedrock), the neutral plane lies at the bedrock surface. For an upward force to be generated on the pile, the pile has to move downward relative to the surrounding soil.

- Due to the solid bedrock, the pile is incapable of moving downward. Hence, the neutral plane would lie at the bedrock surface.

14

Negative Skin Friction and Bitumen-Coated Pile Design

14.1 Introduction

Negative skin friction is the process through which skin friction due to soil acts downward. When skin friction is acting downward, pile capacity decreases.

Negative skin friction (also known as down drag) can be a major problem in some sites.

The causes of negative skin friction are as follows.

- Placement of fill—Fill needs to be placed to gain required elevation for the floor slab. Additional fill would consolidate clay soils underneath. Settling clays would drag the pile down.

- Change in groundwater level—When groundwater level in a site goes down, buoyancy force diminishes. Hence, the effective stress in clay will increase, causing the clay layer to settle. Settling clay layer would generate a negative skin friction on the pile.

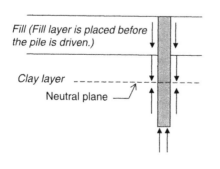

Fill (Fill layer is placed before the pile is driven.)

Clay layer

Neutral plane

- Fill layer is placed on top of the clay layer to achieve necessary elevations.
- A pile is driven into the clay layer.
- Due to weight of the fill, the clay layer will start to consolidate and settle.
- Settlement of the clay layer will produce a down drag on the pile.
- Effective pile capacity could be significantly reduced due to down drag.

Figure 14.1 Negative skin friction due to fill layer

- In most situations, fill has to be placed to achieve necessary elevations. The weight of the fill would consolidate the clay layer.

- Generally, the piles are driven after placing the fill layer.

- Settlement of the clay layer (due to the weight of the fill) induces a down drag force on the pile.

- If the piles are driven after full consolidation of the clay layer has occurred, no down drag will occur.

- Unfortunately, full consolidation may not be completed for years, and many developers may not wait that long to drive piles.

14.2 Bitumen-Coated Pile Installation

- Negative skin friction can be effectively reduced by providing a bitumen coat around the pile.

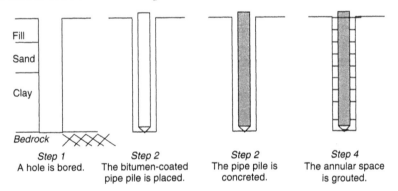

| Step 1 | Step 2 | Step 2 | Step 4 |
| A hole is bored. | The bitumen-coated pipe pile is placed. | The pipe pile is concreted. | The annular space is grouted. |

Figure 14.2 Bitumen coated pile installation

How Bitumen Coating Would Work Against Down Drag

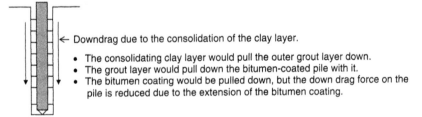

← Downdrag due to the consolidation of the clay layer.

- The consolidating clay layer would pull the outer grout layer down.
- The grout layer would pull down the bitumen-coated pile with it.
- The bitumen coating would be pulled down, but the down drag force on the pile is reduced due to the extension of the bitumen coating.

Figure 14.3 Bitumen coating and the down drag

Typical Situation

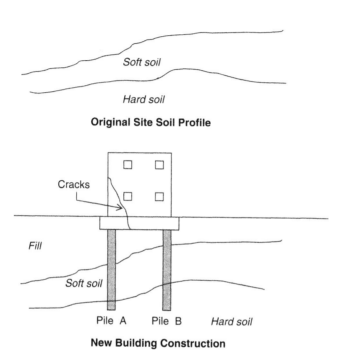

Original Site Soil Profile

New Building Construction

Figure 14.4   New building configuration

Note: Due to larger fill load acting on the left half of the building, negative skin friction acting on pile A will be greater than pile B. Hence, pile A would undergo larger settlement than pile B. Tension cracks could form as shown due to differential settlement.

14.3   Bitumen-Coated Pile Design

Bitumen-coated piles are used to reduce negative skin friction.

Causes of Negative Skin Friction

- Embankment loads
- Groundwater drawdown

Embankment Loads

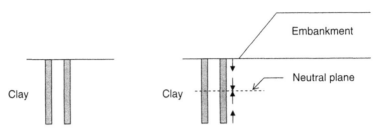

Figure 14.5   Embankment loads

- Negative skin friction occurs above the neutral plane. (See the previous section under "Negative Skin Friction").

- Negative skin friction occurs due to consolidation of the clay layer. When the clay layer settles, it drags the pile down.

Negative Skin Friction Due to Groundwater Drawdown

Figure 14.6   Negative skin friction due to groundwater drawdown

Effective stress at point A (Case 1—prior to drawdown) $= \gamma \cdot x_1 + (\gamma - 62.4) \cdot y_1$

Effective stress at point A (Case 2—after drawdown) $= \gamma \cdot x_2 + (\gamma - 62.4) \cdot y_2$

$\gamma =$ density of soil

   As can be seen, effective stress at point A is larger in case 2 since x_2 is greater than x_1. Higher effective stress would cause the clay to consolidate.

Bitumen Coating

- Bitumen coating should be applied above the neutral plane. In some cases, it may not be necessary to apply bitumen on the full length of the pile above the neutral plane.

- Bitumen can reduce the skin friction by 50 to 90%.

Bitumen Behavior

It is important to understand the behavior of bitumen prior to designing bitumen-coated piles. Bitumen does not behave as a solid or fluid; it has its own behavior.

Shear Strain Rate

Geotechnical engineers are very familiar with shear stress and shear strain. On the other hand, *shear strain rate* is rarely encountered in geotechnical engineering.

Consider a steel cylinder.

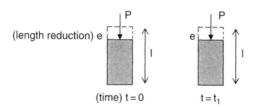

(length reduction) e

(time) t = 0   t = t₁

Figure 14.7 Shear strain

• When a steel cylinder is subjected to a load of P, it undergoes a length reduction of e. After a time period of t_1, length reduction e remains the same. (Creep behavior of steel is ignored.) Stress and strain in steel do not change with time; hence, shear strain rate is zero for solids for all practical purposes.

Bitumen Shear Strain Rate

Figure 14.8 Bitumen shear strain rate

$$t(\text{time}) = 0, \quad \tau = \tau_1, \quad \varepsilon = \varepsilon_1 \quad t = t, \quad \tau = \tau_2, \quad \varepsilon = \varepsilon_2$$
$$\tau = \text{shear stress and } \varepsilon = \text{shear strain}$$

• Assume that at time (t) = 0, shear stress is τ_1 and shear strain is ε_1. At time t = t, the bitumen will deform and the area will change. Hence, shear stress and shear strain will change.

- At t = t, shear stress is τ_2 and shear strain is ε_2.

  Hence, shear strain rate

$$(\gamma) = (\varepsilon_2 - \varepsilon_1)/t \qquad (1)$$

Does the Shear Strain Rate Vary with the Temperature?

- Bitumen deforms faster at high temperatures. Hence, shear strain rate depends on the temperature.

The Shear Strain Rate Is Dependent on Shear Stress

High shear stress produces a high shear strain rate. Hence, shear strain rate is dependent on shear stress as well.

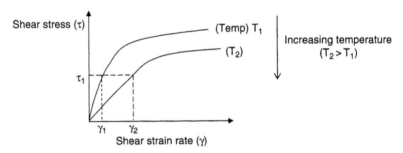

Figure 14.9   Shear stress vs shear strain rate

- When temperature increases from T_1 to T_2, the shear strain rate increases for a given shear stress.

- Similarly, application of a higher shear stress will increase the shear strain rate.

Viscosity

- Viscosity is defined as the ratio of shear stress to shear strain at a given temperature.

- Fluids with high viscosity flow more slowly than fluids with low viscosity. High viscous bitumen flows more slowly than water. Bitumen has a higher viscosity than water.

Viscosity at Temperature (T) = Shear stress/Shear Strain Rate at Temperature (T)

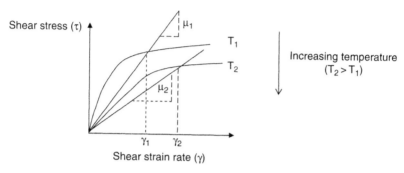

Figure 14.10   Shear stress vs shear strain rate with viscosity

μ_1 = Viscosity of bitumen at temperature T_1 and shear strain rate γ_1
μ_2 = Viscosity of bitumen of temperature T_2 and shear strain rate γ_2

PEN Number

The penetration or PEN number is defined as the penetration of a standard needle into bitumen under a load of 100 g at 25 C in 5 seconds.

Figure 14.11   Bitumen penetration test

- A 100-g standard needle is placed on bitumen for 5 seconds. After 5 seconds, the penetration is measured in tenths of millimeters.

- If the penetration is 1 mm, the PEN number will be 10.

Designing Bitumen-Coated Piles for Negative Skin Friction

STEP 1: Find the mean ground temperature.

Bitumen viscosity and shear strain rate are dependent on the ground temperature.

Mean ground temperature is approximately the same as mean air temperature.

Mean air temperatures (in Fahrenheit) for major U.S. cities and states
are as follows.
NYC 55 F, Upstate NY 40 F, Florida 70 F, LA 60 F, Arizona 70 F, Seattle 50 F
PA 55 F, CO 50 F, U.S. states near the Canadian border 40 F,
States near the Mexican border 70 F (*Visher, 1954*)

STEP 2: Find the required bitumen shear stress (τ).

Typically, bitumen shear stress is selected as 10% of the unit negative
skin friction. In this case, 90% reduction of the negative skin friction
occurs. If this is too costly, then bitumen shear stress could be selected
as 20% of the unit negative skin friction to reduce the negative
skin friction by only 80%. In order to obtain the required bitumen
shear stress, it is necessary to find the unit negative skin friction.
(See Chapters 4 and 5 to compute the unit negative skin friction.)

Simple Formulas to Find Unit Negative Skin Friction

For Clays: Unit negative skin friction $= \alpha \cdot C_u$
$C_u =$ undrained shear strength.
For Sands: Unit negative skin friction $= K \cdot \sigma_v' \cdot \tan (\delta)$

Example

If average unit negative skin friction along the shaft is found to be 100
kN/sq m, then shear stress (τ) in bitumen is selected to be 10 kN/sq m.

STEP 3: Bitumen thickness (d)

Typical bitumen thickness varies from 10 mm to 20 mm.

STEP 4: Settlement rate (SR)

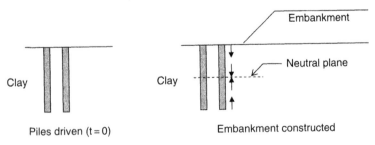

Figure 14.12   Settlement rate

After the embankment is constructed, the largest settlement rate occurs at the start. With time, the settlement rate tapers down.

Find the settlement rate using the consolidation theory during the first month.

$$(1 \text{ month} = 2.6 \times 10^6 \text{s})$$

Settlement rate (first month) (m/s) = Settlement/$(2.6 \times 10^6 \text{ seconds})$

STEP 5: Bitumen viscosity required to control the down drag.

Bitumen viscosity required for down drag (μ_d) is given by the following equation:

$$\mu_d = (\tau \cdot d)/SR \quad (\textit{Briaud, 1997}) \tag{2}$$
$$\tau = \text{shear stress in bitumen (kN/m}^2\text{)},$$
$$d = \text{thickness of the bitumen layer (m)},$$
$$SR = \text{settlement rate (m/s)}$$

STEP 6: Find a bitumen with viscosity less than μ_d.

Inquire from bitumen manufacturers for a suitable bitumen, which would have a viscosity less than computed μ_d. Provide the operating temperature (mean ground temperature) to the Bitumen manufacturer since viscosity changes with temperature.

Why look for a bitumen that has a viscosity less than μ_d?

If the soil settles faster than calculated, the settlement rate increases. Hence, the right-hand side of the above equation would decrease. For this reason, bitumen with viscosity less than μ_d should be selected. When viscosity goes down, fluidity increases.

Design Example

A soil is computed to settle by 0.02 m during the first month of loading. A bitumen thickness of 15 mm was assumed. Unit negative skin friction between soil and pile was computed to be 150 kN/sq m. Ground temperature was found to be 50 F. Find the viscosity required to reduce the negative skin friction by 80%.

Solution

- Negative skin friction = 150 kN/sq m
- Bitumen shear stress (τ) = 150 × 20% = 30 kN/sq m (to obtain 80% reduction)
- Settlement rate during first month = (0.02/2.6 × 10^6) m/s = 7.6 × 10^{-9} m/s
- Bitumen thickness = 15 mm = 0.015 m

$$\mu_d = (\tau \cdot d)/SR \tag{2}$$
$$\mu_d = (30 \times 0.015)/7.6 \times 10^{-9} \text{ kPa} \cdot \text{s} = 6 \times 10^7 \text{ kPa} \cdot \text{s}$$

Specification to the Bitumen Manufacturer Should Include the Following

- Bitumen should have a viscosity less than 6×10^7 kPa · s at 50F and at a shear stress level of 30 kN/sq m.

Bitumen Behavior during Storage

If bitumen is stored at a high temperature environment, it melts and peels off from the pile.

Other than the temperature, another important factor is period of storage. If bitumen is stored for a prolonged period of time, it slowly deforms and peels off from the pile.

STEP 1: Estimate the time period piles need to be stored.

STEP 2: Estimate the mean storage temperature. If piles are to be stored outside, mean air temperature during the storage time period should be estimated.

STEP 3: Find the shear strain rate induced in bitumen coating due to gravity.

$$\gamma = 1/t \tag{3}$$

γ = Shear strain rate induced due to gravity (s^{-1}); t = Storage time period (seconds)

(Briaud, 1997)

STEP 4: Compute the viscosity (μ_s) needed to ensure proper storage using the following equation.

$$\mu_s = (\rho\ t\ d) \tag{4}$$

ρ = density of bitumen measured in $kN/Cu \cdot m$; d = bitumen thickness (m); t = storage time period (in seconds)

(Briaud, 1997)

Design Example

Assume the unit weight of bitumen to be 1,100 Kg/cu m, bitumen thickness to be 0.015 m, storage time to be 20 days, and storage temperature 50F. Find the viscosity required for proper storage purposes.

Solution

$\mu_s = (\rho\ t\ d)$
Density = 1,100 Kg/cu m = 11,000 N/cu m = 11 kN/cu m
$t = 20 \times 24 \times 60 \times 60 = 1.73 \times 10^6$ seconds; d (thickness) = 0.015 m
$\mu_s = \rho\ t\ d = (11) \times (1.73 \times 10^6) \times 0.015 = 2.85 \times 10^5$ K·Pa·s
Induced shear strain rate (from Eq. 3) = $1/(1.73 \times 10^6) = 5.7 \times 10^{-7}$ sec^{-1}

Specification to the Bitumen Manufacturer

Provide a bitumen with a viscosity higher than 2.85×10^5 K·Pa·s at a temperature of 50F at a shear strain rate of 5.7×10^{-7} sec^{-1}.

Note: Bitumen with a viscosity higher than the computed value should be used for storage purposes. When the viscosity goes up, fluidity goes down. Hence, high viscous bitumen is better during storage.

Bitumen Behavior during Driving

It is necessary to make sure that the bitumen will not be damaged during driving. In some cases it may not be possible to save the Bitumen coating during driving. Predrilling a hole is a common solution to avoid damage.

STEP 1: Find the shear strain rate during driving. Equation 3 can be used to find the shear strain rate.

$$\gamma = 1/t \tag{3}$$

It is found that after each blow, bitumen recovers. t is the time period when the pile hammer is in contact with the pile during pile driving. Usually this is a fraction of a second.

STEP 2: Find the viscosity needed to maintain bitumen integrity during driving (μ_{dr}).
Low viscous bitumen is damaged when driving in a high strength soil. (High viscous bitumens are less fluid.) For this reason, high viscous bitumen should be selected when driving in a high-strength soil.

$$\mu_{dr} = (\tau \text{ soil}) \cdot t \quad (\textit{Briaud, 1997}) \tag{5}$$

(τ soil) = Shear strength of soil (kPa)

   For clay soils

(τ soil) = C_u (usually taken as the undrained shear strength)
For sandy soils
(τ soil) = σ' tan ϕ' (σ' = effective stress, ϕ' = friction angle)
Shear strength of sand varies with depth. Hence, average value along the shaft needs to be taken.
t = time period per one blow (seconds)

The viscosity of bitumen should be greater than the computed μ_{dr}.

Temperature

It is assumed that bitumen will be under the same temperature as storage temperature during driving. It is assumed that there is not enough time for the bitumen to reach the ground temperature during driving.

Design Example

Assume that the time period per blow is 0.015 s and that the shear strength of soil is 150 Kpa. The time period per blow is basically the time period when pile hammer would be in contact with the pile. The storage temperature of the pile is 50 F. Find the viscosity requirement for pile driving.

Solution

STEP 1: Find the shear strain rate during driving.

$$\gamma = 1/t \qquad\qquad (3)$$
$$\gamma = 1/0.015 = 66.7 \, \text{sec}^{-1}$$

STEP 2: Find the viscosity needed to maintain bitumen integrity during driving (μ_{dr}).

$$\mu_{dr} = (\tau \text{ soil}) \cdot t$$
$$\mu_{dr} = (150) \times 0.015 = 2.25 \, \text{KPa} \cdot \text{s}$$
$$\text{(Viscosity should be greater than 2.25 Kpa} \cdot \text{s.)}$$

Note: When the viscosity of bitumen goes up, fluidity goes down.

Specification to the Bitumen manufacturer

Provide a bitumen with a viscosity greater than 2.25 K·Pa·s at a temperature of 50 F and at a shear strain rate of 66.7 sec^{-1}.

Final Bitumen Selection

Selected bitumen should comply with all the conditions (conditions for down drag, storage, and driving).

• Bitumen viscosity requirement for negative skin friction control

Provide a bitumen with a viscosity less than 6×10^7 kPa·s at 50 F at a shear stress level of 30 kN/sq m.

- Bitumen viscosity requirement for storage

Provide a bitumen with a viscosity greater than 2.85×10^5 K·Pa·s at a temperature of 50 F at a shear strain rate of 5.7×10^{-7} sec^{-1}.

- Bitumen viscosity requirement for driving

Provide a bitumen with a viscosity greater than 2.25 K·Pa·s at a temperature of 50 F at a shear strain rate of 66.7 sec^{-1}.

This could be done by preparing a bitumen selection chart.

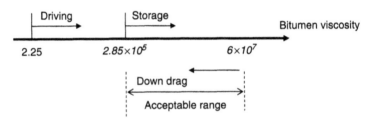

Figure 14.13   Bitumen viscosity

References

Bakholdin, B.V., and Berman, V.I., "Investigation of Negative Skin Friction on Piles," *ASCE J. Soil Mechanics and Foundation Eng.*, 2, no. 4, 238–244, 1975.

Baligh, M.M., et al., "Downdrag on Bitumen Coated Piles," *J. Geotech. Eng., ASCE* 104, no. 11, 1355–1370, 1978.

Briaud, J.L., "Bitumen Election for Reduction of Downdrag on Piles," *ASCE Geotechnical and Geoenvironmental Eng.* Dec. 1997.

Claessen, A.I.M., and Horvat, E., "Reduction of Negative Skin Friction with Bitumen Slip Layers," *J. Geotech. Eng., ASCE* 100, no. 8, 925–944, 1974.

Visher, S.S., Climate Atlas of the United States, Harvard University Press, Cambridge, MA, 1954.

14.4   Case Study: Bitumen-Coated Piles

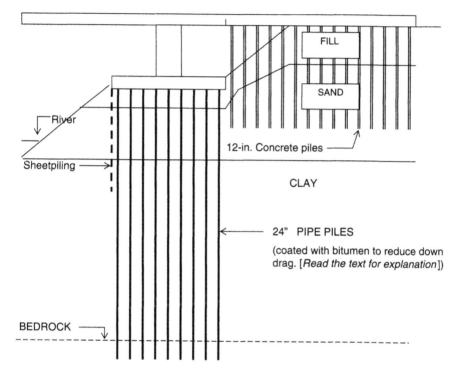

Figure 14.14   Bitumen coated pile case study

24" PIPE PILES: Concrete-filled pipe piles were used to support the abutment. These piles were extended to the bedrock. Initial calculations were done to investigate the possibility of ending the piles in the sand layer. The settlements were found to be too large if they were to be ended in the sand layer. The capacity of the piles was estimated to be 150 tons per pile.

Why Pipe Piles?

The clay layer would undergo settlement due to the fill above. When the clay layer settles, it carries the piles down with it, creating negative skin friction (down drag) on piles. The negative skin friction forces can be as high as 100 tons per pile. The capacity of the piles is not more than 150 tons per pile. The effective capacity (the capacity that is useful) of the pile will be 50 tons per pile. This is not economical.

A bitumen coating (1/8 in. thick) was used to reduce the down drag. Bitumen-coated piles were placed on bored holes. Holes were bored, and the piles were placed inside the hole.

Note: H-piles have less perimeter area compared to similar pipe piles. Hence, H-piles would have less down drag. On the other hand, H-piles are much more expensive than pipe piles (without bitumen coating). Cost comparison between bitumen-coated pipe piles and H-piles was done, and bitumen-coated pipe piles were selected.

15

Pile Design in Expansive Soils

Expansive soils swell when water is introduced. Such soils can create major problems for both shallow foundations and pile foundations.

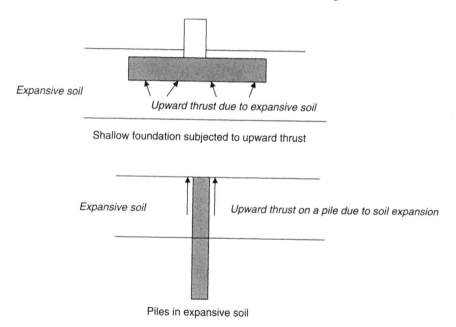

Expansive soil

Upward thrust due to expansive soil

Shallow foundation subjected to upward thrust

Expansive soil

Upward thrust on a pile due to soil expansion

Piles in expansive soil

Figure 15.1 Foundations in expansive soil

Piles in Expansive Soil

Expansive soils are present in many countries, and in some cases, have caused large financial losses.

Identification of Expansive Soils

Expansive soils are mostly silty clays. Not all clays can be considered expansive soils. If the design engineer suspects the presence of expansive soil, he or she should obtain the expansive index of the soil.

The expansive index test is described in ASTM D 4829, "Standard Test Method for Expansion Index of Soils." In this test, a soil specimen is compacted into a metal ring so that the degree of saturation is between 40 and 60%. The specimen is placed in a consolidometer, and a vertical pressure of 1 psi is applied to the specimen. Next, the specimen is inundated with distilled water, and the deformation of the specimen is recorded for 24 hours. The swell or the expansion of the soil volume is computed.

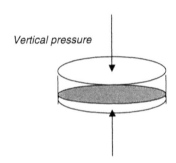

Vertical pressure

Figure 15.2   Expansive soil index test

Expansion Index = Expansion of soil volume/Initial soil volume

Expansive Index	General Guidelines
0 to 20	No special design is needed.
20 to 50	Design engineer should consider effects of expansive soil.
50 to 90	Special design methods to counter expansive soil is needed.
90 to 130	Special design methods to counter expansive soil is needed.
Greater than 130	Special design methods to counter expansive soil is needed.

Pile Design Options

The design engineer should identify expansive soil by conducting expansion index tests. Local codes provide design guidelines under expansive soil conditions. Typically, no action is necessary if the expansion index is less than 20%. If the expansion index is between 20 and 50%, the design engineer can ignore the skin friction developed in expansive soils. If the expansive index is greater than 50%, the pile should be designed against uplift due to expansive soil.

  Solutions for expansive soil conditions

- *Excavate and remove expansive soils* This option is possible only if the expansive soil is limited to a small area of the site.

- *Design against the uplift due to expansive soils* The pile should be able to withstand the uplift caused by expansive soils. The pile should be embedded deep in the soil so that the uplift forces due to expansive soils will be resisted.

- *Ignore the positive skin friction in expansive soil section* The design engineer should ignore any positive skin friction in expansive soils.

Pile Caps

The space between pile caps and the expansive soil strata should be provided for the soil underneath to expand.

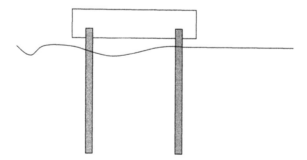

Figure 15.3   Raised foundations

16

Wave Equation Analysis

16.1 Introduction

In 1931, D. V. Isaacs first proposed the wave equation (Smith, 1960). In 1938, E.N. Fox published a solution to the wave equation. During that time, computers did not exist, and Fox used many simplifying assumptions. Smith (1960) solved the wave equation without simplifying assumptions using a computer. According to Smith, his numerical solution was within 5% of the analytical solution. Smith's solution was based on the finite difference technique.

- Prior to discussing the wave equation, it is important to look at dynamic equations.

- Fundamental Driving Formula → $Q \cdot s = W \cdot h$

- Assumptions

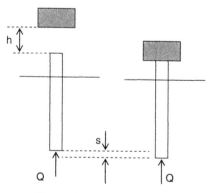

Figure 16.1 Pile driving

It is assumed that Q is a constant force acting at the bottom of the pile. This assumption is not correct. The resistance of the pile is due to two forces: the end bearing force and skin friction. Neither of these forces remains constant during pile movement. (The pile moves by s inches in a downward direction.)

Dynamic equations do not consider stress distribution in the pile, pile diameter, or type of pile. For instance, both piles shown in Figure 16.2 would give the same bearing capacity.

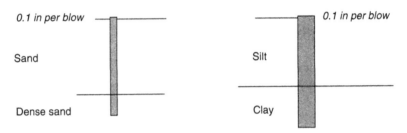

Figure 16.2   Same driving energy for different piles

- The hammer is dropped on the pile, and the set (s) is measured.
- The set is the distance that the pile had moved into the ground.
- h is the drop of the hammer.
- Energy imparted to the pile by the hammer $= W \cdot h$
- Energy used by the pile $= Q \cdot s$
- Energy imparted to the pile is equaled to energy used by the pile. (Energy losses are ignored.)
- Hence, $W \cdot h = Q \cdot s$
- The two piles shown are located in completely different soil conditions. Dynamic equations will not differentiate between the two cases.
- Dynamic equations consider the pile to be rigid during driving. In reality, the pile would recoil and rebound during driving.
- Springs

• Springs

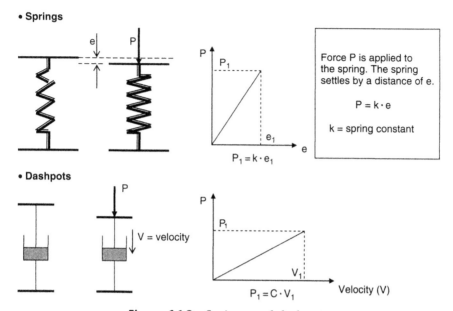

Force P is applied to the spring. The spring settles by a distance of e.

$$P = k \cdot e$$

k = spring constant

• Dashpots

V = velocity

Velocity (V)

Figure 16.3   Springs and dashpots

- In the case of dashpots, the force is proportional to the *velocity* of the dashpot. Above, C is known as the dashpot constant.

Springs → Force is proportional to distance → $P = k \cdot e$
Dashpots → Force is proportional to velocity → $P = C \cdot V$

Representation of Piles in Wave Equation Analysis

- The pile is broken down into small segments.

- Skin friction is represented by springs and dashpots acting on each segment.

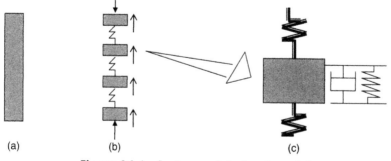

(a)   (b)   (c)

Figure 16.4   Spring and dashpot model

These figures are as follows:

A pile is shown in Figure 16.4a.

In Figure 16.4b the pile is divided into segments. If the pile is divided into more segments, the accuracy of the analysis can be improved. On the other hand, it would take more computer time for the analysis.

One segment is shown in Figure 16.4c. The skin friction is represented with a dashpot and a spring.

Wave Equation

$$\frac{\partial^2 u}{\partial t^2} = c^2 \frac{\partial^2 u}{\partial x^2}$$

u = displacement; c = wave speed; t = time; x = length

The finite difference method is used to solve the above equation. Computer programs such as WEAP and GRLWEAP are available in the market to perform the wave equation analysis.

16.2   Soil Strength under Rapid Loading

This chapter is designed to provide information regarding the strength properties of soil under rapid loading situations. During pile driving, the pile is subjected to rapid loading.

Experiments have shown that failure stress is higher under rapid loading conditions than during gradual loading.

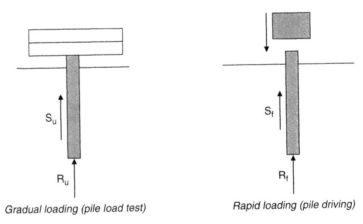

Gradual loading (pile load test)   Rapid loading (pile driving)

Figure 16.5   Gradual loading and rapid loading

- During pile driving, the pile is subjected to rapid loading.

   R_u = ultimate tip resistance during *static* loading
   R_f = ultimate tip resistance during *rapid* loading
   S_u = ultimate skin friction during *static* loading
   S_f = ultimate skin friction during *rapid* loading

Equation for Tip Resistance for Rapid Loading Condition

The following relationship has been developed for tip resistance under rapid loading condition.

$$R_f = R_u\left(1 + J_u \times V^{Nu}\right)$$

S_u = ultimate skin friction during *static* loading
S_f = ultimate skin friction during *rapid* loading
J_u = rate effect parameter for tip resistance
V = velocity of the pile
N_u = rate effect exponent for tip resistance
J_u = $1.2 - 0.007\ C_u$ (for clay soils)

(*Lee et al., 1988*)

C_u = (cohesion measured in kN/sq m)
Units of J_u would be $(\text{sec/m})^N$
$J_u = 2.2 - 0.08\ (\phi - 20)$ (for sandy soils)

(*Lee et al., 1988*)

ϕ = Friction angle of sand
Units of R_f and R_u = kN/sq m

(*Note: The above equations were developed by the author based on experimental data provided by Lee et al., 1988.*)

   N_u => Experiments show that N_u lies between 0.17 and 0.37 for both sandy and clayey soils. Hence, Nu can be taken as 0.2.

Equations for Skin Friction for Rapid Loading Condition

The following relationship has been developed for skin friction under rapid loading condition.

$$S_f = S_u\left(1 + J_s \times V^{N_s}\right)$$

S_u = ultimate skin friction during *static* loading
S_f = ultimate skin friction during *rapid* loading
J_s = rate effect parameter for skin friction
V = velocity of the pile
N_s = rate effect exponent for skin friction
J_s = 1.6 – 0.008 C_u (for clay soils)

(Lee et al., 1988)

C_u = (cohesion measured in kN/sq m)
Units of Js would be $(sec/m)^N$
J_s = 0 (for sandy soils)

(Heerema, 1979)

N_s => Experiments show that Ns lies between 0.17 and 0.37 for both sandy and clayey soils. Hence, N_s can be taken as 0.2 (Heerema, 1979). Both N_s (skin friction exponent) and N_u (tip resistance exponent) both lie in the same range.
Units of S_f and S_u = kN/sq m

References

Coyle, H.M., and Gibson, G.C., "Empirical Damping Constants for Sands and Clays," *ASCE J. of Soil Mechanics and Foundation Eng.*, 949–965, 1970.

Dayal, V., and Allen, J.H., "The Effect of Penetration Rate on The Strength of Remolded Clay and Sand," *Canadian Geotechnical Eng. J.*, 12, no. 3, 336–348, 1975.

Heerema, E.P., "Relationships Between Wall Friction Displacement Velocity and Horizontal Stress in Clay and in Sand for Pile Driveability Analysis," *Ground Engineering*, 12, no. 1, 55–60, 1979.

Lee, S.L., et al. "Rational Wave Equation Model for Pile Driving Analysis," *ASCE Geotechnical Eng. J.*, March 1988.

Litkouhi, S., and Poskitt, T.J., "Damping Constants for Pile Driveability Calculations," *Geotechnique*, 30, no. 1, 77–86, 1980.

16.3 Wave Equation Analysis Software

- Geotechnical engineers are required to provide information to wave equation computer programs for analysis.

Example of Input Data for Wave Equation Software

Pile Hammer Data

- Hammer type—Delmag D 12-32 Diesel Hammer (single acting)
- Hammer energy—31,320 ft lbs
- Hammer efficiency—80%
- Blows per minute—36
- Hammer weight (striking part only)—2,820 lbs
- Hammer stroke—11'1"
  (The engineer should obtain the equivalent stroke for double-acting hammers from the manufacturer.)
- Hammer efficiency—80%

Capblock Data

- Type—Micarta sheets
- Modulus of elasticity—70,000 psi
- Coefficient of restitution—0.6
- Diameter—12 in.
- Thickness—6 in.

Pile Cushion Data

Usually, pile cushions are used only for concrete piles.
   If a pile cushion is used, material, modulus of elasticity, and dimensions of the cushion should be provided.

Pile Properties

- Pile material—steel
- Density—475 lbs/cu ft
- Outer diameter of pile—24 in.
- Wall thickness—¼ in.

- Modulus of elasticity—30,000 psi

- Pile is driven closed end.

- Pile embedment—30 ft

Note: For concrete piles, the area of steel reinforcements and prestress force also should be provided.

Soil Information

Depth 0–1—Top soil	SPT (N) values—(10,13)
Depth 1–5—Silty sand	SPT (N) values—(15,13), (12,18), (13,15), (19,10)
Depth 5–9—Soft clay	SPT (N) values—(2,2), (3,1), (2,5), (4,7)

- Cohesion of clay (C_u) (Cohesion values should be provided for clay soils.)

- Rapid loading parameters for soil layers (J_u, J_s, N_u and N_s)

(Note: See Chapter 15 for an explanation of these parameters.)

- Density of soil

- G_s (shear modulus of soil). G_s can be obtained experimentally. If not, the following approximate equations can be used. See under "Shear Modulus."

  $G_s = 150\ C_u$ (for clays); Gs would have same units as C_u.
  $G_s = 200\ \sigma_v'$ (for sands); $\sigma v' = $ Vertical effective stress.

(Lee et al., 1988)

- Information provided by wave equation programs

Wave equation programs are capable of providing the following information.

- Blows required to penetrate a certain soil stratum (or rate of penetration per blow)

- Pile capacity

- Ability of the given pile hammer to complete the project on a timely basis.

References

Lee, S.L., et al. "Rational Wave Equation Model for Pile Driving Analysis," *ASCE Geotechnical Eng. J.*, March 1988.
Smith, E.A.L. (1960), "Pile Driving Analysis by Wave Equation," *ASCE Soil Mechanics and Foundation Eng.*, 35–50, August 1960.

Companies

GRLWEAP: Pile Dynamics Inc., Ohio. Tel: 216-831-6131,
Fax: 216-831-0916

17

Batter Piles

Batter piles are used for bridge abutments, retaining walls, and platforms.

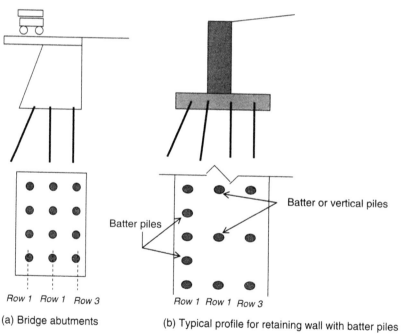

(a) Bridge abutments

(b) Typical profile for retaining wall with batter piles

Figure 17.1   Batter piles

- In most cases, one row of piles is battered as in Figure 17.1a. In some cases it is necessary to have more than one row of batter piles, as in Figure 17.1b.

Theory: Forces on Batter Piles

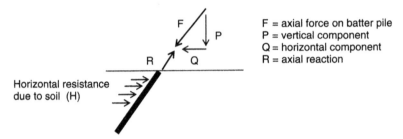

F = axial force on batter pile
P = vertical component
Q = horizontal component
R = axial reaction

Figure 17.2   Forces on a batter pile

Batter piles are capable of resisting significant lateral forces. Lateral resistance of batter piles comes from two sources.

1. Horizontal component of the axial reaction

2. Horizontal resistance due to soil (H)

Note: Horizontal resistance (H) and horizontal component of the pile axial reaction are two different quantities.

Negative Skin Friction

Batter piles should be avoided in situations where negative skin frictional forces can be present. Settling soil could induce large bending moments in batter piles.

Thrust Due to Settling Soil

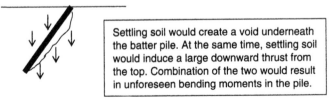

Settling soil would create a void underneath the batter pile. At the same time, settling soil would induce a large downward thrust from the top. Combination of the two would result in unforeseen bending moments in the pile.

Figure 17.3   Batter pile in a settling soil

Force Polygon for Figure 17.1a

Please refer to Figure 17.1a for the following force polygon. (Calculation of loads will be shown later in the chapter.)

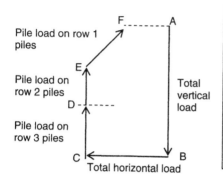

Line AB: Calculate the total vertical load on piles. Draw the total vertical load.
Line BC: Draw the total horizontal load (load due to wind or water).
Piles should be able to withstand these two loads.
Line CD: Draw the load on piles along row 3. These piles are vertical.
Line DE: Draw the load on piles along row 2.
Line EF: Draw the load on piles along row 1. These are batter piles.
Line FA: This line indicates the lateral force required to stabilize the piles. Lateral resistance of pile should be more than the load indicated by line FA.

Figure 17.4   Force diagram

Force Polygon for Figure 17.1b

Please refer to Figure 17.1b for the following force polygon.

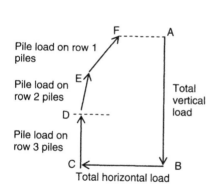

Line AB: Calculate the total vertical load on piles. Draw the total vertical load.

Line BC: Draw the total horizontal load.

Line CD: Draw the load on piles along row 3. These piles are vertical.
Line DE: Draw the load on piles along row 2. These piles are batter piles.
Line EF: Draw the load on piles along row 1.
Line FA: This line indicates the lateral force required to stabilize the piles. Lateral resistance of the pile should be more than the load indicated by line FA.

Figure 17.5   Force diagram

Design Example 1

Compute the loads on piles due to the retaining wall shown. Assume the lateral earth pressure coefficient at rest (K_0) to be 0.5. Piles in the front are battered at $20°$, and center piles are battered at $15°$ to the vertical.

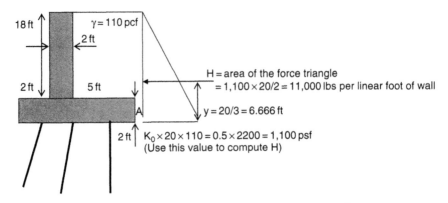

Figure 17.6 Batter piles in a retaining wall

STEP 1: Horizontal force

Horizontal force H acts 6.666 ft from the base. ($y = 20/3$)
Moment (M) $= 11,000 \times 6.666$ lb/ft $= 73,333$ lb/ft per linear ft of wall
Moment in 3-linear-foot section $= 3 \times 73,333 = 219,999$ lb/ft
Assume the piles have been placed at 3-ft intervals.

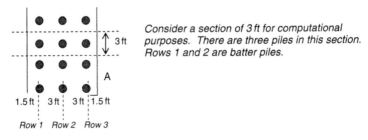

Figure 17.7 Batter piles – plan view

STEP 2: Weight of soil and concrete

Weight of soil resting on the retaining wall (height $= 18$, width $= 5$)
$= 5 \times 18 \times 110 = 9,900$ lbs
Weight of concrete $= (2 \times 9 + 2 \times 18) \times 160 = 8,640$ lbs (160 pcf $=$ concrete density)
Total weight $= 18,540$ lbs per linear feet of wall
Total weight (W) in a 3-ft section $= 55,620$ lbs

There are three piles in the 3-ft section. Hence, each pile carries a vertical load of 18,540 lbs.

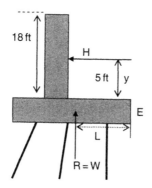

Figure 17.8   Batter piles – elevation view

STEP 3: Overturning Moment

Overturning moment = Resisting moment
Take moments around point E
$R \times L = W \times L = M = H \cdot y = 219{,}999$ lb/ft   (See step 1.)
$L = 219{,}999/W = 219{,}999/55{,}620 = 3.96$ ft
$M = $ overturning moment due to H

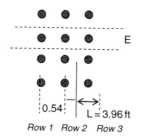

Figure 17.9   Batter piles – plan view

STEP 4: Stability

The center of gravity of the piles lies along the center of the middle row. Vertical reaction acts 3.96 ft from the edge.

Distance to the reaction from the center of the pile system = $4.5 - 3.96$ = 0.54 ft

Moment around the centerline = $R \times 0.54 = 55{,}620 \times 0.54 = 30{,}038.4$ lbs per 3-ft section

Note that the moment around the centerline is different from the moment around the edge (point E).

STEP 5: Additional load due to bending moments:

- Each pile carries a vertical load of 18,540 lbs. (See step 2.)

- Due to the moment, the base of the retaining wall undergoes a bending moment. This bending moment can create additional stress on piles. Stress developed on piles due to bending is given by the following equation.

$$\frac{M}{I} = \frac{\sigma}{y}$$

M = bending moments; σ = bending stress; I = moments of area; y = distance

Find the moment of area of piles by taking moments around the centerline of the footing.
Moment of area of row 1 = A × 3^2 = 9A   (A = area of piles)
Moment of area of row 2 = A × 0 = 0   (Distance is taken from the center of gravity of piles.)
Moment of area of row 3 = A × 3^2 = 9A
   Total moment of area = 18 A

$$\text{Bending load on piles in row 1 } (\sigma) = \frac{M}{I} \cdot y = \frac{30,034.8}{18\,A} \times 3$$
$$= \frac{5005.8}{A} \text{ lbs/sq ft}$$

M = 30,034.8 lbs per 3-ft section (See step 4)
y = 3 ft (distance from centerline to corner row of piles)

$$\text{Bending load per pile} = \frac{5005.8}{A} \times A = 5005.8 \text{ lbs per pile}$$

STEP 6: Total loads on piles

Total vertical load on pile = Vertical load + Load due to bending
Total vertical load on piles in row 1 = 18,540 + 5005.8 = 23,545.8 lbs
Bending load is positive for piles in row 1.
Total vertical load on piles in row 2 = 18,540 + 0 = 18,540 lbs   (since y = 0)
Load due to bending is zero for row 2.

Total vertical load on piles in row 3 = 18,540 − 5005.8 = 13,534.2 lbs
Bending load is negative for piles in row 3.

STEP 7: Force polygon

Draw the force polygon for the 3-ft section selected. Total horizontal
load for a 3-ft section is 33,000 lbs (11,000 lbs per linear foot; see step 1).
Total vertical load for a 3-ft section is 55,620 lbs; see step 4. Start
drawing the force polygon from I to J. HI is the final leg.

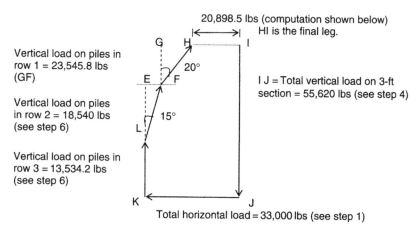

Figure 17.10   Force distribution diagram

$EF = 18,540 \times \tan 15 = 4451$ lbs; $GH = 23,545.8 \tan 20 = 7,650.5$ lbs
Total horizontal load resisted = 4,451 + 7,650.5 = 12,101.5 lbs
Additional load that needs to be resisted = 33,000 − 12,101.5 =
   20,898.5 lbs
This load needs to be resisted by the soil pressure acting on the piles
   laterally. (See Figure 17.11.)

The lateral pile resistance should not be confused with axial pile
resistance. In the case of a vertical pile, axial load is vertical, and there
is no horizontal component to the axial load.

On the other hand, when a horizontal load is applied to a vertical
pile, it will resist. Hence, all piles (including vertical piles) have a
lateral resistance.

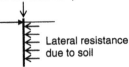

Figure 17.11   Vertical pile

Figure 17.12   Batter pile

Axial load P can be broken down into X and Y components. Computation of lateral resistance of the pile due to soil pressure is given in the chapter 11.

There are three piles in the considered 3-ft section. Each pile should be able to resist a load of 6,966.5 lbs. Use the principles given in "Lateral Pile Resistance" to compute the lateral pile capacity due to soil pressure.

A factor of safety (FOS) of 2.5 to 3.0 should be used. For instance, required vertical pile capacity for piles in row 1 is 23,545 lbs. The piles should have an ultimate pile capacity of 70,635 lbs assuming an FOS of 3.0.

Note: No credit should be given to the resistance due to soil friction acting on the base of the retaining wall. Piles would stress the soil due to the horizontal load acting on them. Stressed soil underneath the retaining wall will not be able to provide any frictional resistance to the retaining wall. If piles were to fail, piles would pull the soil with them. Hence, soil would not be able to provide any frictional resistance to the retaining structure.

Design Example 2

Compute the loads on piles due to the retaining wall shown. Assume the lateral earth pressure coefficient at rest (K_0) to be 0.5. Piles in the front are battered at 20°, and center piles are battered at 15° to the vertical. (This problem is similar to the previous example, except for the pile configuration.)

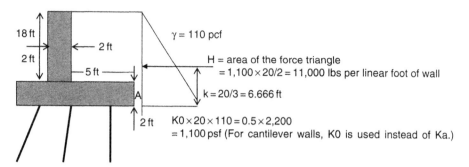

Figure 17.13   Load distribution in a batter pile—elevation view

STEP 1: Horizontal force H, acts 6.666 ft from the base.

Moment (M) = 11,000 × 6.666 lb/ft = 73,333 lb/ft per linear ft of wall
Moment in 6-linear-foot section = 6 × 73,333 = 439,998 lb/ft
Assume the piles have been placed at 3-ft intervals.

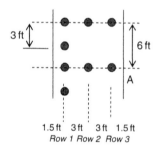

Figure 17.14   Batter piles—plan view

Consider a section of 6 ft for computational purposes. There are six half piles and one full pile in this section. Rows 1 and 2 are batter piles. Row 1 has twice as many piles than rows 2 and 3.

Weight of soil resting on the retaining wall = 5 × 18 × 110 = 9,900 lbs
Weight of concrete = (2 × 9 + 2 × 18) × 160 = 8,640 lbs   (160 pcf = concrete density)
Total weight = 18,540 lbs per linear feet of wall
Total weight (W) in a 6-ft section = 111,240 lbs

There are four piles in the 6-ft section. Hence, each pile carries a load of 27,810 lbs.

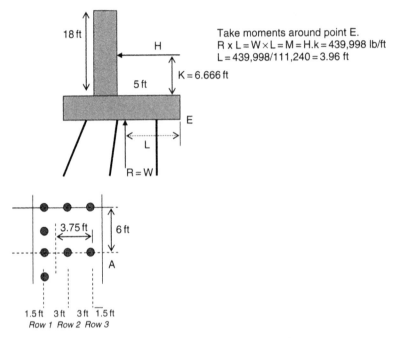

Take moments around point E.
$R \times L = W \times L = M = H.k = 439,998$ lb/ft
$L = 439,998/111,240 = 3.96$ ft

Figure 17.15   Batter pile example

Center of Gravity of Piles

There are six half piles and one full pile in this section.
There are two half piles and one full pile in row 1.
There are two half piles in row 2 and two half piles in row 3. (Total piles in the section = 4)
Take moments around row 3.
$(2 \times 6 + 1 \times 3)/4 = 3.75$ ft

- As per the above calculations, the center of gravity of piles is located 3.75 ft from row 3.
- The moment of area of the system has to be computed from the center of gravity of piles.

  Distance to row 1 from center of gravity = $(6 - 3.75)$ ft = 2.25 ft
  Distance to row 2 from center of gravity = 0.75 ft
  Distance to row 3 from center of gravity = 3.75 ft

STEP 2: Each pile carries a load of 27,810 lbs. (See step 1.)

• Stress developed on piles

$$\frac{M}{I} = \frac{\sigma}{y}$$

M = bending moment; σ = bending stress; I = moment of area;
y = distance

Find the moment of area of piles.

Row 1

Moment of area of row 1 = $A \cdot r^2$ = $A \times 2.25^2$ = 10.12 A   (A = area of piles)
Moment of area of row 2 = $A \times 0.75^2$ = 0.56 A
(Distance is taken from the center of gravity.)
Moment of area of row 3 = $A \times 3.75^2$ = 14.06 A
Total moment of area = 24.74 A
Bending moment (M) = 439,998 lb/ft (See step 1)

$$\text{Bending load on piles in row 1 } (\sigma) = \frac{M}{I} \cdot y = \frac{439,998}{24.74\,A} \times 2.25$$
$$= \frac{40,016}{A} \text{ lbs}$$

$$\text{Bending load on row 1 piles} = \frac{40,016}{A} \times 2A = 80,032 \text{ lbs}$$

(There are two piles in the first row. Two half piles and one full pile is
   equivalent to 2 piles.)
Total load on pile = Vertical load + Load due to bending
Total load on piles in row 1 = 27,810 × 2 + 80,032 = 135,652 lbs
Vertical load per pile is 27,810 lbs. Since there are two piles in row 1,
   multiply 27,810 lbs by 2.

STEP 3: Row 2

$$\text{Bending load on piles in row 2 } (\sigma) = \frac{M}{I} \cdot y = \frac{439,998}{24.74\,A}$$
$$\times 0.75 = \frac{13,339}{A} \text{ lbs}$$

$$\text{Bending load per pile} = \frac{13{,}339}{A} \times A = 13{,}339 \text{ lbs per pile}$$

(There are two half piles in row 2, which amounts to one pile.)
Multiply by area of the pile to convert the load from lbs per sq ft to lbs per pile.

$$\text{Total load on piles in row } 2 = 27{,}810 - 13{,}339 = 14{,}471 \text{ lbs}$$

(Since center of gravity is on the opposite side, the bending moment creates a tensile force on the pile.)

STEP 4: Row 3

$$\text{Bending load on piles in row 3 } (\sigma) = \frac{M}{I} \cdot y = \frac{439{,}998}{24.74\,A}$$

$$\times 3.75 = \frac{66{,}693}{A} \text{ lbs}$$

$$\text{Bending load per pile} = \frac{66{,}693}{A} \times A = 66{,}693 \text{ lbs per pile}$$

Row 3 has two half piles, which amounts to one pile.

Total load on piles in row $3 = 27{,}810 - 66{,}693 = -38{,}883$ lbs per pile

Above, 27,810 lbs is the vertical load on piles, and 66,693 lbs is the load due to bending, which is tensile.

The total force on piles in row 3 is tensile.

STEP 5: Draw the force polygon.

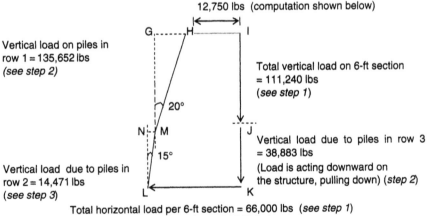

12,750 lbs (computation shown below)

Vertical load on piles in
row 1 = 135,652 lbs
(see step 2)

Total vertical load on 6-ft section
= 111,240 lbs
(see step 1)

20°

Vertical load due to piles in row 3
= 38,883 lbs
(Load is acting downward on
the structure, pulling down) (step 2)

15°

Vertical load due to piles in
row 2 = 14,471 lbs
(see step 3)

Total horizontal load per 6-ft section = 66,000 lbs (see step 1)
(horizontal force per linear foot = 11,000 lbs)

Figure 17.16   Load distribution diagram

$$NM = 14,471 \times \tan 15 = 3,877 \text{ lbs;}$$
$$GH = 135,652 \times \tan 20 = 49,373 \text{ lbs} \quad \text{(This load accounts for two piles in row 1.)}$$

Total horizontal load resisted by axial forces of piles $= 49,373 + 3,877 = 53,250$ lbs

Additional load that needs to be resisted $= 66,000 - 53,250 = 12,750$ lbs

This load needs to be resisted by the soil pressure acting on piles laterally.

Each pile should have a lateral pile capacity of $12,750/4$ lbs $= 3,188$ lbs

18

Vibratory Hammers — Design of Piles

18.1 Introduction

Although many piles have been driven in clay soils using vibratory hammers, vibratory pile hammers are best suited for sandy soils. Vibratory hammers are less noisy and do not cause pile damage as compared to pile driving hammers.

One of the major problems of vibratory hammers is the unavailability of credible methods to compute the bearing capacity of piles based on penetration rates. On the other hand, pile-driving formulas can be used to compute the ultimate bearing capacity of piles driven with drop hammers.

Vibratory Pile-Driving Depends on:

- Pile and soil characteristics
- Elastic modulus of pile and soil (E_p and E_s)
- Lateral earth pressure coefficient (K_0)
- Relative density of soil (D_r)
- Vibratory hammer properties

18.2   Vibratory Hammer Properties

The effectiveness of a vibratory hammer depends on:

- Frequency of the vibratory hammer (Frequency = Number of vibrations per second)
- Amplitude of the hammer (distance traveled during up and down motion of the vibratory hammer)
- Vibrating weight of the hammer
- Surcharge weight of the hammer (In addition to the vibrating weight, a nonvibrating weight is also attached to the hammer to produce a downward thrust. Bias weight is another name for the surcharge weight).

Definitions

- Eccentric moment = Rotating weight (m) × Eccentricity (e)
- Centrifugal force = $C \cdot m \cdot e \cdot \omega^2$

m = Eccentric weight or the rotating weight. This is different from the vibrating weight (M). Vibrations are caused by the rotating weight. The rotating weight is attached to a housing, and the whole unit is allowed to vibrate due to the rotating weight.

- Vibrating weight = Rotating weight + Bias weight + Weight of the housing
- C = constant depending on the vibratory hammer

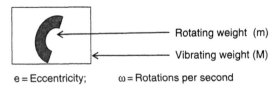

e = Eccentricity;   ω = Rotations per second

Figure 18.1   Vibratory hammer weights

- Eccentric moment = m · e
- A (amplitude) = distance travel during up and down motion of the vibrating weight.

Some parameters of vibratory pile hammers are given below as an example.

Vibratory Pile Hammer Data (ICE—Model 416L)

Eccentric moment (m · e) = 2,200 lbs. in = 2,534 Kg. cm
Frequency (ω) = 1,600 vpm (vibrations per minute)
Centrifugal force = 80 tons = 712 kN

The pile hammer manufacturer (ICE) provides the following equation to compute the centrifugal force for their hammers:

Units (U.S. tons)

Centrifugal force = $0.00005112 \times (m \cdot e \cdot \omega^2)$ tons
m should be in lbs, e should be in inches, and ω should be in vibrations per second.

Units (kN)

Centrifugal force = $0.04032 \times (m \cdot e \cdot \omega^2)$ kN
m should be in Kgs, e should be in meters, and ω should be in vibrations per second.

Design Example

Find the centrifugal force of a vibratory pile hammer, which has an eccentric weight (m) of 1,000 Kgs, an eccentricity (e) of 2 cm, and a vibrating frequency (ω) of 30 vibrations per second.

Solution

$$\text{Centrifugal force} = 0.04032 \times (m \cdot e \cdot \omega^2) \text{ kN}$$
$$= 0.04032 \times (1,000 \times 0.02 \times 900) \text{ kN} = 725 \text{ kN}$$

18.3   Ultimate Pile Capacity

The ultimate capacity of a pile driven using a vibrating hammer can be obtained by using the equation proposed by Feng and Deschamps (2000).

$$Q_u = \frac{3.6\ (F_c + 11 \cdot W_b)}{\left[1 + V_p\ (OCR)^{\frac{1}{2}}\right]}\ \frac{L_e}{L}$$

Q_u = ultimate pile capacity (kN); F_c = centrifugal force (kN):
W_b = surcharge weight or bias weight (kN); V_p = penetration velocity (m/min)
OCR = overconsolidation ratio; L_e = embedded length of the pile (m)
L = total length of the pile (m)

Design Example

A vibratory hammer has the following properties.
Surcharge weight of the hammer = 8 kN;   Centrifugal force = 600 kN
Find the penetration rate (V_p) required to obtain an ultimate pile capacity of 800 kN. Assume the pile to be fully embedded and the OCR of the soil to be 1.0.

Solution

$$Q_u = \frac{3.6\ (F_c + 11 \cdot W_b)}{\left[1 + V_p\ (OCR)^{\frac{1}{2}}\right]}\ \frac{L_e}{L}$$

$$800 = \frac{3.6\ (600 + 11 \times 8)}{\left[1 + V_p\ (1)^{\frac{1}{2}}\right]}\ \frac{L_e}{L}$$

($L_e = L$, since the pile is fully embedded)
$V_p = 2.096$ m/min
(The pile should penetrate 2.096 m/min or less to achieve an ultimate pile capacity of 800 kN.)

Penetration Rate and Other Parameters

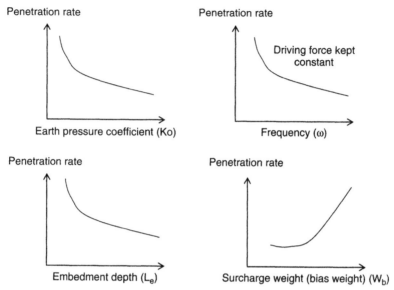

Figure 18.2   Penetration rates

Vibratory Hammers (International Building Code—IBC):

IBC states that vibratory drivers shall be used only when pile load capacity is verified by pile load tests.

References

Bernhard, R.K., "Pile Soil Interactions During Vibro Pile Driving," *ASTM J. of Materials* 3, No. 1, March 1968.

Davisson, M.T., "BRD Vibratory Driving Formula," *Foundation Facts*, 1990, 6, No. 1, 9–11.

Feng, Z., and Deschamp, R.J. "A Study of the Factors Influencing the Penetration and Capacity of Vibratory Driven Piles," *Soils and Foundations*, Japanese Geotechnical Society, June 2000.

O' Neill, M.W., Vipulanandan, C., and Wong, D.O., "Laboratory Modeling of Vibro Driven Piles," *ASCE, J. of Geotechnical Eng.* 116, No. 8, 1990.

Rodger, A.A., and Littlejohn, G.S., "A Study of Vibratory Driving on Granular Soils," *Geotechnique* 30, No. 3, 269–293, 1995.

19

Seismic Analysis of Piles

19.1 A Short Course on Seismology

A general understanding of seismology is needed to design piles for seismic events.

Figure 19.1a shows a pendulum moving back and forth drawing a straight line prior to an earthquake event. Figure 19.1b shows the same pendulum drawing a wavy line during an earthquake event.

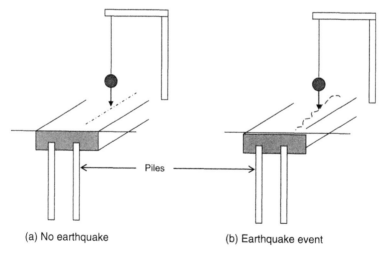

(a) No earthquake   (b) Earthquake event

Figure 19.1 Earthquake measurement

- Seismographs are designed using the above principle.

- Due to the movement of the pile cap, piles are subjected to additional shear forces and bending moments.

- Earthquakes occur as a result of disturbances occurring inside the earth's crust. Earthquakes produce three main types of waves.

  1. P-waves (primary waves): P-waves are also known as compressional waves or longitudinal waves.

  2. S-waves (secondary waves): S-waves are also known as shear waves or transverse waves.

  3. Surface waves: Surface waves are shear waves that travel near the surface.

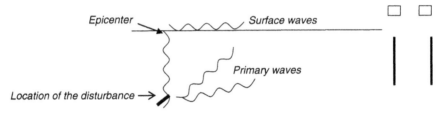

Figure 19.2   Seismic waves

Faults

Faults—fractures where a block of earth has moved relative to the other—are a common occurrence in the earth. Fortunately, most faults are inactive and will not cause earthquakes.

Horizontal Fault

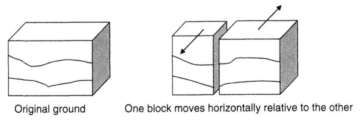

Figure 19.3   Movement near faults

- In a horizontal fault, one earth block moves horizontally relative to the other.

Vertical Fault (Strike Slip Faults)

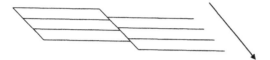

Figure 19.4   Vertical movement

In this type of fault, one block moves in a downward direction relative to the other.

Active Fault

An active fault is defined as a fault in which there was an average historic slip rate of 1 mm/year or more during the last 11,000 years (*IBC*).

19.1.1   Richter Magnitude Scale (M)

$$M = Log\ (A) - Log\ (A_0) = Log\ (A/A_0)$$

M = Richter magnitude scale; A = maximum trace amplitude during the earthquake;

A_0 = standard amplitude (A standard value of 0.001 mm is used for comparison. This corresponds to a very small earthquake.)

Design Example 1

What is the Richter magnitude scale for an earthquake that recorded an amplitude of (a) 0.001 mm, (b) 0.01 mm, and (c) 10 mm?

Answer

(a) $A = 0.001$ mm

$M = Log\ (A/A_0) = Log\ (0.001/0.001) = Log\ (1) = 0$
$M = 0$

(b) $A = 0.01$

$M = Log\ (A/A0) = Log\ (0.01/0.001) = Log\ (10) = 1$
$M = 1$

(c) A = 10 mm

$$M = Log (A/A_0) = Log (10/0.001) = Log (10,000) = 4$$
$$M = 4$$

Largest Earthquakes Recorded

Location	Date	Magnitude
Chile	1960	9.5
Prince William, Alaska	1964	9.2
Aleutian Islands	1957	9.1
Kamchatka	1952	9.0
Ecuador	1906	8.8
Rat Islands	1965	8.7
India-China Border	1950	8.6
Kamchatka	1923	8.5
Indonesia	1938	8.5
Kuril Islands	1963	8.5

Source: USGS Web site (*www.usgs.org* United States Geological Survey).

19.1.2 Peak Ground Acceleration

This is a very important parameter for geotechnical engineers. During an earthquake, soil particles accelerate. Acceleration of soil particles can be either horizontal or vertical.

19.1.3 Seismic Waves

The following partial differential equation represents seismic waves.

$$G(\partial^2 u/\partial x^2 + \partial^2 u/\partial z^2) = \rho(\partial^2 u/\partial t^2 + \partial^2 v/\partial t^2)$$

G = shear modulus of soil; ∂ = partial differential operator;
u = horizontal motion of soil;   v = vertical motion of soil;
ρ = density of soil;

Fortunately, geotechnical engineers are not called upon to solve this partial differential equation.

A seismic wave form described by the above equation would create shear forces and bending moments on piles.

Seismic Wave Velocities

The velocity of seismic waves is dependent on the soil/rock type. Seismic waves travel much faster in sound rock than in soils.

Soil/Rock Type	Velocity (ft/sec)
Dry Silt, sand, loose gravel, loam, loose rock, moist fine-grained top soil	600–2500
Compact till, gravel below water table, compact clayey gravel, cemented sand, sandy clay	2,500–7,500
Weathered rock, partly decomposed rock, fractured rock	2,000–10,000
Sound shale	2,500–11,000
Sound sandstone	5,000–14,000
Sound limestone and chalk	6,000–20,000
Sound igneous rock (granite, diabase)	12,000–20,000
Sound metamorphic rock	10,000–16,000

Source: Peck, Hanson, and Thornburn (1974).

References

IBC, International Building Code.
Peck, R.B., Hanson, W.B., and Thornburn, T.H., *Foundation Engineering*, John Wiley and Sons, New York, 1974.
USGS Web site, United States Geologic Survey, www.usgs.org

19.2 Seismic Pile Design

Earthquakes can cause additional bending moments and shear forces on piles. Earthquake-induced bending moments and shear forces can be categorized into three types.

1. Kinematic loads

2. Inertial loads

3. Loads due to liquefaction

19.2.1 Kinematic Loads

Seismic waves travel at *different velocities in different soils.* Because of these differences, piles are subjected to bending and shear forces. This bending is known as kinematic pile bending.

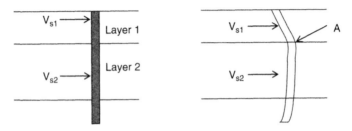

Figure 19.5 Pile deformation due to earthquakes

V_{s1} and V_{s2} are seismic wave velocities in layers 1 and 2, respectively. The pile is subjected to differential forces due to different seismic waves arriving in two soil layers. Kinematic pile bending can occur in homogeneous soils as well, owing to the fact that seismic wave forms can have different strengths depending on the depth and surrounding structures that could damp the wave in a nonuniform manner.

- Kinematic pile bending can also occur in free piles (i.e., piles that are not supporting building structures).

- The maximum bending moment in the pile occurs near the interface of two layers. In Figure 19.5, maximum bending moment occurs at point A.

19.2.2 Inertial Loads

In addition to kinematic loads, seismic waves can induce inertial loads.

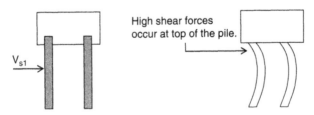

Figure 19.6 Earthquake loading on piles

Inertial Loading Mechanism

The inertial loading mechanism on piles is different from the kinematic loading mechanism. Inertial loading occurs due to building mass acting on piles.

When a seismic wave reaches the pile, the pile is accelerated. Assume the acceleration at the top of the pile to be "a". The pile is not free to accelerate because it is attached to a pile cap and then to a structure on top. Due to the mass of the structure on top, the pile will be subjected to inertial forces and bending moments.

Shear force at top of the pile can be computed using Newton's equation:

$$\text{Shear force} = M \times a$$

Where M = portion of the building mass acting on the pile
     a = acceleration of the pile

- Inertial loadings due to seismic waves are limited to the top 10 d to 15 d measured from the surface.
  (d = diameter of the pile)

- On the other hand, kinematic loading can occur at any depth. If a pile fails at a greater depth, then it is reasonable to assume that the pile has failed due to kinematic loading.

19.2.3   Soil Liquefaction

Soil liquefaction occurs in sandy soils below the groundwater level. During an earthquake, loose sandy soils tend to liquefy (i.e., reach a liquid-like state). When this happens, the piles lose the lateral support provided by soil.

19.3   Design of Piles for Kinematic Loadings

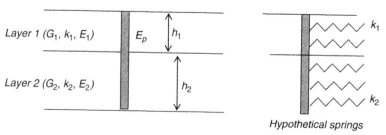

Figure 19.7   Spring model

G = soil shear modulus
k = soil subgrade modulus or the spring constant
E = Young's modulus of soil
E_p = Young's modulus of pile

Relationships

$$k = 3G$$

(Dobry & O'Rourke, 1983)

The (k/G) ratio ranges from 2.5 to 4.0 for soils. k = 3G is a good approximation for any soil. The pile bending moment is proportional to $(k)^{1/4}$, and any attempt to find an accurate value for k is not warranted.

$$k_1/E1 = k_2/E_2 = \delta$$

(Mylanakis, 2001)

E1, E2 = Young's modulus of soil layers 1 and 2, respectively
Here δ is a dimensionless parameter. δ can be computed using the
  following equation.

$$\delta = 3/(1 - v^2)[(E_p/E_1)^{-1/8}(L/d)^{1/8}(h_1/h_2)^{1/2}(G^2/G_1)^{-1/30}]$$

d = pile diameter; L = pile length; E_p = Young's modulus of the pile;
  h_1, h_2 = thickness of soil layers 1 and 2, respectively

Pile Bending Strain

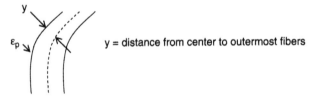

y = distance from center to outermost fibers

Figure 19.8   Pile bending strain

ε_p = Bending strain at the outermost fiber of the pile
E_p (Young's Modulus) = Stress (σ_p)/Strain (ε_p)

$$E = \sigma_p/\varepsilon_p$$

Bending Equation

$$M/I = \sigma_p/y$$

I = moment of inertia, M = bending moment, and y (see Figure 19.8)

$$M = (\sigma_p \cdot l)/y = (E_p \cdot \varepsilon_p \cdot l)/y$$

If ε_p (bending strain) can be deduced, the maximum bending moment in the pile can be computed.

Seismic Pile Design for Kinematic Loads

STEP 1: Find the peak soil shear strain at the soil interface of two layers.

(Seed and Idriss, 1982)

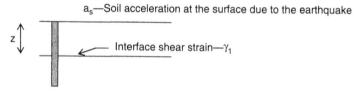

a$_s$—Soil acceleration at the surface due to the earthquake

Interface shear strain—γ_1

Figure 19.9   Interface shear strain

The interface peak shear strain due to the soil acceleration is given below.

$$\gamma_1 = (r_d \times \rho_1 \times h_1 \times a_s)/G_1$$

γ_1 = peak shear strain at the interface of two soil layers
a_s = soil acceleration due to the earthquake at the surface
r_d = depth factor; $r_d = 1 - 0.015\,z$
(z = depth to the interface measured in meters)
ρ_1 = soil density of the top layer
h_1 = thickness of the top soil layer

STEP 2: Find δ (the ratio between k and E)

$$\delta = k_1/E_1 = k_2/E_2$$

k_1 and k_2 = spring constants of layers 1 and 2
E_1 and E_2 = Young's modulus of layers 1 and 2

δ is given by the following equation.

$$\delta = 3/(1 - v^2)[(E_p/E_1)^{-1/8}(L/d)^{1/8}(h_1/h_2)^{1/2}(G^2/G_1)^{-1/30}]$$

STEP 3: Find the strain transfer ratio (ε_p/γ_1).

$$(\varepsilon_p/\gamma_1) = \frac{(c^2 - c + 1)\{[3(k_1/E_p)^{1/4}(h_1/d) - 1]\,c(c - 1) - 1\}}{2c^4(h_1/d)}$$

$c = (G_2/G_1)^{1/4}$
ε_p = bending strain at the outermost fiber of the pile
γ_1 = interface shear strain (computed in step 1)

Values of G_1, G_2, k_1, k_2, E_p, h_1, h_2 and d are required to compute ε_p.

STEP 4: Find the bending moment (M) in the pile.

$$M = (\sigma_p \cdot I)/y$$

(See the earlier section, "Pile Bending Strain")

$$E \text{ (Young's modulus)} = \text{Stress/Strain} = \sigma_p/\varepsilon_p$$

$$M = (E_p \cdot \varepsilon_p \cdot I)/y$$

Since ε_p was calculated and y, I, and E_p are known pile properties, the bending moment (M) induced due to earthquake can be deduced.

Design Example 2

Find the kinematic bending moment induced in the pile shown due to an earthquake that produces a surface acceleration of 0.5 g.
    Young's modulus of pile material $(E_p) = 3.5 \times 10^7$ Kpa; pile diameter $= 0.5$ m

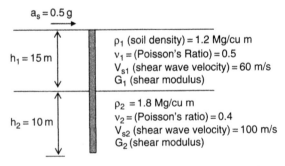

Figure 19.10   Seismic pile design example

The following parameters need to be deduced from given information (G_1, G_2, E_1, ε_p).

Note that instead of soil shear modulus, shear wave velocity is given.

STEP 1: Find G_2/G_1 (shear modulus ratio of two layers).

Shear modulus is proportional to soil density and shear wave velocity. The relationship between shear wave velocity and soil shear modulus is as follows.

$G = \rho \times V_s^2$

Hence, $G_1 = \rho_1 \times V_{s1}^2$ and $G_2 = \rho_2 \times V_{s2}^2$

$G_1 = 1.2 \times 60^2 = 4320$; $G_2 = 1.8 \times 100^2$

$$G_2/G_1 = (1.8 \times 100^2)/(1.2 \times 60^2) = 4.17$$

STEP 2: Find the Young's modulus of soil layer 1.

The following relationship between Young's modulus and shear modulus can be used to find the Young's modulus of the soil layer.

$E = 2G(1 + v)$

$E_1 = 2G_1(1 + v_1) = 2 \times 4320\,(1 + 0.5) = 12{,}960\,\text{kPa}$

STEP 3: Find the ratio (E_p/E_1)

$E_p/E_1 = (3.5 \times 10^7)/(12{,}960) = 2{,}700$

(E_p—Young's modulus of the pile is given, and E_1 is computed in step 2.)

STEP 4: Find the peak soil shear strain at the soil interface of two layers.

(Seed and Idriss, 1982)

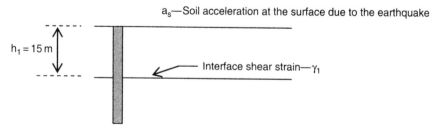

Figure 19.11   Interface shear strain and soil acceleration

The peak interface shear strain due to the soil acceleration follows.

$$\gamma_1 = (r_d \times \rho_1 \times h_1 \times a_s)/G_1$$

r_d = depth factor; $r_d = 1 - 0.015\ z$ (z = depth to the interface measured in meters)

$r_d = 1 - 0.015 \times h1 = 1 - 0.015 \times 15 = 0.775$

ρ_1 (soil density) $= 1.2\ \text{Mg/cu m}$, $a_s = 0.5$ g; $G_1 = 4320$ (computed in step 1)

$\gamma_1 = (0.775 \times 1.2 \times 15 \times 0.5 \times 9.81)/4320 = 1.6 \times 10^{-2}$

STEP 5: Find δ (The ratio between k and E).

$$\delta = k_1/E_1 = k_2/E_2$$

k_1 and k_2 = spring constants in layers 1 and 2
E_1 and E_2 = Young's modulus in layers 1 and 2

δ is given by the following equation:

$$\delta = \{3/(1 - v^2)\} \times [(E_p/E_1)^{-1/8}(L/d)^{1/8}(h_1/h_2)^{1/2}(G_2/G_1)^{-1/30}]$$

$E_p/E_1 = 2{,}700$ (calculated in step 3); $G_2/G_1 = 4.17$ (calculated in step 1); Pile diameter (d) $= 0.5\ \text{m}$ (pile diameter is given); L = pile length $= 25\ \text{m}$

$\delta = \{3/(1 - 0.5^2)\} \times [(2700)^{-1/8}(25/0.5)^{1/8}(15/10)^{1/2}(4.17)^{-1/30}]$

$\delta = 4 \times [0.372 \times 1.63 \times 1.22 \times 0.954] = 2.82$

STEP 6: Find the strain transfer ratio (ε_p/γ_1).

$$(\varepsilon_p/\gamma_1) = \frac{(c^2 - c + 1)\{[3(k_1/E_p)^{1/4}(h_1/d) - 1]\,c(c - 1) - 1\}}{2c^4(h_1/d)}$$

$c = (G_2/G_1)^{1/4} = 1.429$
$G_2/G_1 = 4.17$ (see step 1)
ε_p = bending strain at the outermost fiber of the pile
γ_1 = interface shear strain (see step 4)

All parameters in the above equation are known except for the soil spring constant (k_1). The soil spring constant can be computed using the following equation:

$$\delta = k_1/E_1 = k_2/E_2$$

$E_1 = 12{,}960$ Kpa (see step 2); $\delta = 2.82$ (from step 5).

$2.82 = k_1/12{,}960$; $k_1 = 36{,}547$ Kpa; $E_p = 3.5 \times 10^7$ Kpa (pile parameter)

$$(\varepsilon_p/\gamma_1) = \frac{(c^2 - c + 1)\{[3(k_1/E_p)^{1/4}(h1/d) - 1]c(c - 1) - 1\}}{2c^4(h_1/d)}$$

$$\varepsilon_p/\gamma_1 = \frac{(1.429^2 - 1.429 + 1)\{[3[36{,}547/(3.5 \times 10^7)]^{1/4}}{(15/0.5) - 1] \times 1.429 \times (1.429 - 1) - 1\}}{2 \times 1.429^4(15/0.5)}$$

$$\varepsilon_p/\gamma_1 = \frac{1.613\{[16.17 - 1] \times 0.613 - 1\}}{250.2} = 0.056.$$

$$\gamma_1 = 1.6 \times 10^{-2} \text{ (from step 4)}$$

Hence, $\varepsilon_p = 1.6 \times 10^{-2} \times 0.056 = 8.96 \times 10^{-4}$

STEP 7: Find the bending moment (M) in the pile.

$$M = (\sigma_p\,I)/y$$

(See the section "Pile Bending Strain.")

E_p (Young's modulus of the pile) = Stress/ Strain = σ_p/ε_p

$$M = (E_p \cdot \varepsilon_p I)/y \quad y = d/2 = 0.5/2 = 0.25 \text{ m}$$

I (moment of inertia) $= \pi \cdot d^4/64 = \pi \cdot 0.5^4/64 = 3.068 \times 10^{-3}$

$$\varepsilon_p = 8.96 \times 10^{-4} \text{ (see step 6)}$$

$$M = \{3.5 \times 10^7 \times 8.96 \times 10^{-4} \times 3.068 \times 10^{-3}\}/0.25 = 384.8 \text{ kN} \cdot \text{m}$$

- The pile should be designed to withstand an additional bending moment of 384.8 kN·m induced by earthquake loading.

19.4   Seismic Pile Design—Inertial Loads

As mentioned earlier, inertial loads occur in piles due to shaking of the building.

Assume a pile as shown in Figure 19.12.

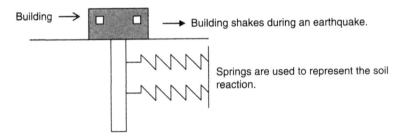

Building shakes during an earthquake.

Springs are used to represent the soil reaction.

Figure 19.12   Spring model for seismic forces

- When the building starts to shake as a result of an earthquake event, the piles are subjected to additional shear forces and bending moments.

- Usually, loads occurring in the piles due to building loads are limited to 15 d from the ground surface (d = diameter of the pile).

- Unfortunately, there is no easy equation to represent the motion of the building and piles. The following partial differential equation has been proposed.

$$\partial V / \partial Z \cdot dZ + \rho \cdot \partial^2 y / \partial t^2 \cdot dZ + k \cdot D(y - u) \cdot dZ = 0$$

V = shear force in the pile at a depth of "Z"; Z = depth; D = pile diameter;

u = horizontal soil movement due to the earthquake; v = horizontal pile movement due to the earthquake

(u − v) = horizontal pile movement relative to soil due to the earthquake; ρ = pile density

- This equation cannot be solved with reasonable accuracy by manual methods. Hence, computer programs are used to calculate the shear forces that develop in piles.

- The SHAKE computer program by Schnable Engineering is one such program that can be used for the purpose.

References

Dobry, R., and O'Rourke, M.J., "Discussion—Seismic Response of End Bearing Piles," *ASCE J. of Geotechnical Eng.*, May 1983.

Dobry, R., and Gazetas, G., "Simple Method for Dynamic Stiffness and Damping of Floating Pile Groups," *Geotechnique* 38, no. 4, 557–574, 1988.

Dobry, R., et al. "Horizontal Stiffness and Damping of Single Piles," *ASCE J. of Geotechnical. Eng.*, 108, no. 3, 439–459, 1982.

Florres-Berrones, R., "Seismic Response of End Bearing Piles," *ASCE J. of Geotechnical Eng.*, April, 1982.

Kavvadas, M., and Gazetas, G., "Kinematic Seismic Response and Bending of Freehead Piles in Layered Soil," *Geotechnique* 43, no. 2, 207–222, 1993.

Kaynia, A.M., and Mahzooni, S., "Forces in Pile Foundations Under Seismic Loading," *ASCE J. of Eng. Mech.* 122, no. 1, 46–53, 1996.

Mylanakis, G., "Simplified Model for Seismic Pile Bending at Soil Layer Interfaces," *Soils and Foundations*, Japanese Geotechnical Society, August 2001.

Nikolaou, S., et al., "Kinematic Pile Bending During Earthquakes; Analysis and Field Measurements," *Geotechnique* 51, no. 5, 425–440, 2001.

Novak, M., "Dynamic Stiffness and Damping of Piles," *Canadian Geotechnical Eng. J.* 2, No. 4, 574–598, 1974.

Poulos, H.G., "Behavior of Laterally Loaded Piles," *ASCE J. of Soil Mechanics and Foundation Eng.*, May 1971.

19.5   Liquefaction Analysis

Theory

Sandy and silty soils tend to lose strength and turn into a *liquid*-like state during an earthquake. This happens in response to the increased pore pressure during an earthquake event in the soil caused by seismic waves.

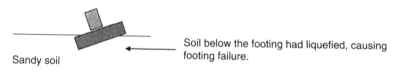

Soil below the footing had liquefied, causing footing failure.

Sandy soil

Figure 19.13   Soil liquefaction

- Liquefaction of the soil was thoroughly studied by Bolten Seed and I.M Idris during the 1970s. As one would expect, liquefaction behavior of soil cannot be expressed in one simple equation. Many correlations and semi-empirical equations have been introduced by researchers. For this reason, Professor Robert W. Whitman convened a workshop in 1985 (Liquefaction Resistance of Soils) on behalf of the National Research Council (NRC). Experts from many countries participated in this workshop, and a procedure was developed to evaluate the liquefaction behavior of soils.

- Only sandy and silty soils tend to liquefy. Clay soils do not undergo liquefaction.

Impact Due to Earthquakes

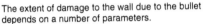

Imagine a bullet hitting a wall.

The extent of damage to the wall due to the bullet depends on a number of parameters.

Bullet Properties
(1) Velocity of the bullet
(2) Weight of the bullet
(3) Hardness of the bullet material

Wall Properties
(1) Hardness of the wall material
(2) Type of wall material

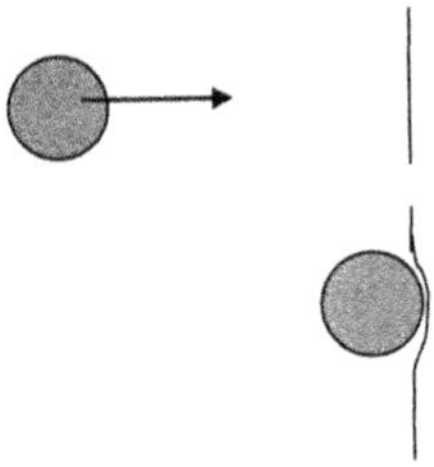

Figure 19.14   Bullet properties

Parameters that affect liquefaction are shown below.

Earthquake Properties
- Magnitude of the earthquake

- Peak horizontal acceleration at the ground surface (a_{max})

Soil Properties
- Soil strength, measured by Standard Penetration Test (SPT) value

- Effective stress at the point of liquefaction

- Content of fines (Fines are defined as particles that pass through the #200 sieve.)

- Earthquake properties that affect soil liquefaction are amalgamated into one parameter known as cyclic stress ratio (CSR).

$$\text{Cyclic stress ratio (CSR)} = 0.65(a_{max}/g) \times (\sigma/\sigma') \times r_d \qquad (1)$$

a_{max} = peak horizontal acceleration at the ground surface
σ = total stress at the point of concern; σ' = effective stress at the point of concern
r_d = stress reduction coefficient (This parameter accounts for the flexibility of the soil profile.)

$$r_d = 1.0 - 0.00765\,Z \quad \text{for} \quad Z < 9.15\,\text{m} \qquad (1.1)$$
$$r_d = 1.174 - 0.0267\,Z \quad \text{for} \quad 9.15\,\text{m} < Z < 23\,\text{m} \qquad (1.2)$$

(Z = depth to the point of concern in meters)

Soil Resistance to Liquefaction

- As a rule of thumb, any soil that has a SPT value higher than 30 will not liquefy.

- As mentioned earlier, resistance to liquefaction of a soil depends on its strength measured by SPT value. Researchers have found that resistance to liquefaction of a soil depends on the content of *fines* as well.

The following equation can be used for a clean sand. (Clean sand is defined as a sand with less than 5% fines.)

$$\text{CRR}_{7.5} = \frac{1}{[34 - (N1)_{60}]} + \frac{(N1)_{60}}{135} + \frac{50}{[10 \cdot (N1)_{60} + 45]^2} - \frac{2}{200} \quad (2)$$

$\text{CRR}_{7.5}$ = soil resistance to liquefaction for an earthquake of 7.5 Richter magnitude.

The correction factor needs to be applied to any other magnitude. That process is described later in the chapter. The above equation can be used only for sands. (Content of fines should be less than 5%.) The correction factor has to be used for soils with a higher content of fines. That procedure is also described later in the chapter.

$(N1)_{60}$ = Standard penetration value corrected to a 60% hammer and an overburden pressure of 100 Kpa. (*Note that the above equations were developed in metric units.*)

How to obtain $(N1)_{60}$

$$(N1)_{60} = Nm \times C_N \times C_E \times C_B \times C_R \quad (3)$$

Nm = SPT value measured in the field
C_N = overburden correction factor = $(Pa/\sigma')^{0.5}$
Pa = 100 Kpa; σ' = Effective stress of soil at point of measurement
C_E = energy correction factor for the SPT hammer
For donut hammers $C_E = 0.5$ to 1.0; for trip type donut hammers
   C_E = 0.8 to 1.3
C_B = borehole diameter correction
For boreholes 65 mm to 115 mm, use $C_B = 1.0$
For a borehole diameter of 150 mm, use $C_B = 1.05$
For a borehole diameter of 200 mm, use $C_B = 1.15$
C_R = Rod length correction (rods attached to the SPT spoon exert their weight on the soil. Longer rods exert a higher load on soil, and in some cases the spoon goes down due to the weight of rods without

any hammer blows. Hence correction is made to account for the weight of rods).

For rod length <3 m use $C_R = 0.75$; for rod length 3 m to 4 m use $C_R = 0.8$; for rod length 4 m to 6 m use $C_R = 0.85$; for rod length 6 m to 10 m use $C_R = 0.95$; for rod length 10 m to 30 m use $C_R = 1.0$

Design Example 3

Consider a point at a depth of 5 m in a sandy soil (fines $<5\%$). Total density of soil is 1,800 Kg/m^3. The groundwater is at a depth of 2 m. The corrected $(N1)_{60}$ value is 15. Peak horizontal acceleration at the ground surface (a_{max}) was found to be 0.15 g for an earthquake of magnitude 7.5. Check to see whether the soil at a depth of 5 m would liquefy under an earthquake of 7.5 magnitude.

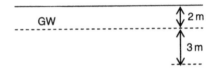

Figure 19.15   Groundwater level

STEP 1: Find the cyclic stress ratio.

$$\text{Cyclic stress ratio (CSR)} = 0.65(a_{max}/g) \times (\sigma/\sigma') \times r_d \quad (1)$$

a_{max} = peak horizontal acceleration at the ground surface = 0.15 g
σ = total stress at the point of concern; σ' = effective stress at the point of concern
r_d = stress reduction coefficient (This parameter accounts for the flexibility of the soil profile.)

$$r_d = 1.0 - 0.00765\,Z \quad \text{for} \quad Z < 9.15\,\text{m} \quad (1.1)$$

(Z = depth to the point of concern in meters)

$$r_d = 1.174 - 0.0267\,Z \quad \text{for} \quad 9.15\,\text{m} < Z < 23\,\text{m} \quad (1.2)$$

$\sigma = 5 \times 1{,}800 = 9{,}000\,\text{Kg/m}^2$
$\sigma' = 2 \times 1{,}800 + 3\,(1{,}800 - 1{,}000) = 6000\,\text{Kg/m}^2$
(Density of water = 1,000 Kgm^2)

Since the depth of concern is 5 m (which is less than 9.15 m), use the above first equation to find r_d.

$$r_d = 1.0 - 0.00765\,Z \quad \text{for} \quad Z < 9.15\,\text{m}; \; r_d = 1.0 - 0.00765 \times 5 = 0.962$$

$$\text{Hence, CSR} = 0.65 \times (0.15) \times \frac{9{,}000}{6{,}000} \times 0.962 = 0.1407$$

STEP 2: Find the soil resistance to liquefaction

$$\text{CRR}_{7.5} = \frac{1}{[34 - (\text{N1})_{60}]} + \frac{(\text{N1})_{60}}{135} + \frac{50}{[10 \cdot (\text{N1})_{60} + 45]^2} - \frac{2}{200} \quad (2)$$

$(\text{N1})_{60}$ value is given to be 15. Hence, $\text{CRR}_{7.5} = 0.155$.

Since soil resistance to liquefaction (0.155) is larger than the CSR value (0.1407), the soil at 5 m depth will not undergo liquefaction for an earthquake of magnitude 7.5.

Correction Factor for Magnitude

As you are aware, Eq. (2) (for $\text{CRR}_{7.5}$) is valid only for earthquakes of magnitude 7.5. The correction factor is proposed to account for magnitudes different than 7.5.

Factor of safety (FOS) is given by $= (\text{CRR}_{7.5}/\text{CSR})$
$\text{CRR}_{7.5} =$ resistance to soil liquefaction for an earthquake of 7.5 magnitude and
$\text{CSR} =$ cyclic stress ratio (which is a measure of the impact due to the earthquake load)

The FOS for any other earthquake is given by the following equation.

$$\text{FOS} = (\text{CRR}_{7.5}/\text{CSR}) \times \text{MSF} \quad (4)$$

$\text{MSF} =$ magnitude scaling factor
MSF is given by the following table.

Table 19.1   Magnitude scaling factors

Earthquake Magnitude	MSF Suggested by Idris (1995)	MSF Suggested by Andrus and Stokoe (1997)
5.5	2.2	2.8
6.0	1.76	2.1
6.5	1.44	1.6
7.0	1.19	1.25
7.5	1.00	1.00
8.0	0.84	
8.5	0.72	

The 1985 NRC conference participants gave engineers the freedom to select either of the values suggested by Idris or Andrus and Stokoe. Idris's values are more conservative, and in noncritical buildings such as warehouses, the engineers may be able to use Andrus and Stokoe values.

Design Example 4

The $CRR_{7.5}$ value of a soil was found to be 0.11. The CSR value for the soil was computed to be 0.16. Will this soil liquefy for an earthquake of 6.5 magnitude?

STEP 1: Find the factor of safety.

$$FOS = (CRR_{7.5}/CSR) \times MSF \qquad (4)$$

MSF (magnitude scaling factor) for an earthquake of 6.5 = 1.44 (Idris)
$FOS = (0.11/0.16) \times 1.44 = 0.99$ (Soil will liquefy.)
Use MSF given by Andrus and Stokoe.
$FOS = (0.11/0.16) \times 1.6 = 1.1$ (Soil will not liquefy.)

Correction Factor for Content of Fines

Equation (2) was developed for clean sand with fines content less than 5%. The correction factor is suggested for soils with higher fines content.

$$CRR_{7.5} = \frac{1}{[34 - (N1)_{60}]} + \frac{(N1)_{60}}{135} + \frac{50}{[10 \cdot (N1)_{60} + 45]^2} - \frac{2}{200} \qquad (2)$$

Corrected $(N1)_{60}$ value should be used in Eq. (2) for soils with higher fines content.

The following procedure should be used to find the correction factor.

- Compute $(N1)_{60}$ as in the previous case.

- Use the following equations to account for the fines content.

$(N_1)_{60C} = a + b(N_1)_{60}$ $(N_1)_{60C} =$ corrected $(N_1)_{60}$ value

$a = 0$ for   FC < 5% (FC = fines content)   (5.1)

$a = \exp[1.76 - 190/FC^2]$ for   5% < FC < 35%   (5.2)

$a = 5.0$ for   FC > 35%   (5.3)

$b = 1.0$ for   FC < 5%   (5.4)

$b = [0.99 + (FC^{1.5}/1,000)]$ for   5% < FC < 35%   (5.5)

$b = 1.2$ for   FC > 35%   (5.6)

Design Example 5

$(N1)_{60}$ value for soil with 30% fines content was found to be 20. Find the corrected $(N_1)_{60C}$ value for that soil.

STEP 1: $(N_1)_{60C} = a + b(N1)_{60}$

For FC = 30%; $a = \exp[1.76 - 190/FC^2]$

$= \exp[1.76 - 190/30^2] = 4.706$ (5.2)

For FC = 30% $b = [0.99 + (FC^{1.5}/1,000)] = 1.154$ (5.5)

$(N1)_{60C} = 4.706 + 1.154 \times 20 = 27.78$

Design Example 6

Consider a point at a depth of 5 m in a sandy soil (fines = 40%). Total density of soil is 1,800 Kg/m³. The groundwater is at a depth of 2 m (γw = 1,000 Kg/m³). Corrected $(N1)_{60}$ value is 15. (All the correction parameters C_N, C_E, C_B, C_R are applied, except for the fines content.) Peak horizontal acceleration at the ground surface (a_{max}) was found to be 0.15 g for an earthquake of magnitude 8.5. Check to see whether the soil at a depth of 5 m would liquefy under this earthquake load.

STEP 1: Find the cyclic stress ratio.

Cyclic stress ratio (CSR) $= 0.65(a_{max}/g) \times (\sigma/\sigma') \times r_d$ (1)

$\sigma = 5 \times 1,800 = 9,000$ Kg/m²; $\sigma' = 2 \times 1,800 + 3(1,800 - 1,000)$

$= 6,000$ Kg/m²

Since the depth of concern is 5 m (which is less than 9.15 m) use Eq. (1.1)

$$r_d = 1.0 - 0.00765\ Z \quad \text{for} \quad Z < 9.15\ \text{m}; \ r_d = 1.0 - 0.00765 \times 5 = 0.962$$

$$\text{Hence, CSR} = 0.65 \times (0.15) \times \frac{9{,}000}{6{,}000} \times 0.962 = 0.1407$$

STEP 2: Provide the correction factor for fines content.

For soils with 40% fines → a = 5 and b = 1.2 (Eqs. 5.3 and 5.6).
$(N_1)_{60C} = a + b\ (N_1)_{60}$
$(N_1)_{60C} = 5 + 1.2 \times 15 = 23$

STEP 3: Find the soil resistance to liquefaction.

$$\text{CRR}_{7.5} = \frac{1}{[34 - (N_1)_{60}]} + \frac{(N_1)_{60}}{135} + \frac{50}{[10 \cdot (N_1)_{60} + 45]^2} - \frac{2}{200} \quad (2)$$

$(N_1)_{60C}$ value is found to be 23. (See step 2.)
Hence, $\text{CRR}_{7.5} = 0.25$ (from Eq. 2)

$$\text{FOS} = (\text{CRR}_{7.5}/\text{CSR}) \times \text{MSF} \quad (4)$$

MSF = magnitude scaling factor

Obtain MSF from Table 19.1.
MSF = 0.72 for an earthquake of 8.5 magnitude (table by Idris)
CSR = 0.1407 (see step 1)
FOS = (0.25/0.1407) × 0.72 = 1.27
The soil would not liquefy.

References

Andrus R.D. and Stokoe, K.H., "Liquefaction Resistance of Soils from Shear Wave Velocity," *ASCE J. of Geotechnical and Geoenvironmental Engineering*, 126, No. 11, 1015–1025, 2000.

Idris, I.M. "Evaluating Seismic Risk in Engineering Practice," *Proc., 11th Int. Conf. on Soil Mech. and Found. Eng*, 255–320, 1985.

Idris, I.M. "Response of soft soil sites during earthquakes," *Proc. Bolten Seed Memorial Symp.*, 2, Bi-Tech Publishers Ltd., Vancouver, 273–290, 1990.

Liquefaction Resistance of Soils, 1996 NCEER and 1998 NCEER/NSF workshop on evaluation of liquefaction resistance of soils. *ASCE J. of Geotechnical and Geoenvironmental Engineering*, 127, No. 10, October 2001.

19.6   General Guidelines for Seismic Pile Design

IBC classifies sites based on SPT value and cohesion values.

Site Class

Site Class	Soil/Rock Type	SPT	Cu (psf)	V_s (Soil Shear Velocity)
A	Hard rock	N/A	N/A	$V_s > 5,000$
B	Rock	N/A	N/A	$2,500 < V_s < 5,000$
C	Very dense soil and soft rock	$N > 50$	$C_u > 2,000$	$1,200 < V_s < 2,500$
D	Stiff soil	$15 < N < 50$	$1,000 < C_u < 2,000$	$600 < V_s < 1,200$
E	Soft soil	$N < 15$	$C_u < 1,000$	$V_s < 600$
F	Peats, liquefiable soils, and high-plastic soils (PI > 75%)			

Pre Cast Concrete Piles

Class C Sites

- Longitudinal reinforcements shall be provided with a minimum steel ratio of 0.01. Lateral ties should be not less than ¼ in. thick.

- Lateral ties should not be placed more than 6 in. apart.

- Longitudinal and lateral reinforcements should be provided for the full length of the pile.

Class D, E and F

- Class C, requirements must be met.

- In addition, lateral ties should not be placed more than 4 in. apart.

20

Pile Design Software

20.1 Introduction

Most geotechnical engineering software are based on the finite element method, which is considered to be the most powerful mathematical method today for solving piling problems.

- Any type of soil condition can be simulated using the finite element method.

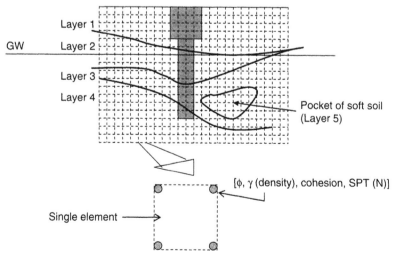

Figure 20.1 Finite element grid

- A complicated soil profile is shown in the figure. Nodes in each of the finite elements are given the soil properties of that layer such as ϕ, γ (density), cohesion, and SPT (N) value.

- Nodes of elements in layer 1 are given the soil properties of layer 1. Similarly, nodes on layer 2 will be given the soil properties of layer 2.

- Because of this flexibility, isolated soil pockets also can be effectively represented.

20.1.1   Representation of Time History

- The capacity of a pile is dependent on the history of loading. A pile that was loaded gradually would have a higher capacity than a pile that was loaded rapidly.

- Assume that a developer is planning to construct a 10-story building in five years. In this case, a full building load on piles would gradually develop in a time period of ten years.

- On the other hand, the developer would change his mind and decide to construct the ten-story building in two years. In this case, full load on piles would develop in two years. If the piles were to be fully loaded in two years, the capacity of piles would be less than in the first scenario.

- In such situations, the finite element method could be used to simulate the time history of loading.

  Groundwater Changes: Change in groundwater conditions also affects the capacity of piles. Change of groundwater level can easily be simulated by the finite element method.

  Disadvantages: The main disadvantage of the finite element method is its complex nature. In many cases, engineers may wonder whether it is profitable to perform a finite element analysis.

Finite Element Computer Programs

Computer programs are available with finite element platforms. These programs can be used to solve a wide array of piling problems. The user is expected to have a working knowledge of finite element analysis to use these programs. More specialized computer programs are also available in the market. These programs do not require knowledge of finite element analysis.

20.2   Boundary Element Method

- The boundary element method, a simplified version of the finite element method, considers only the elements at boundaries.

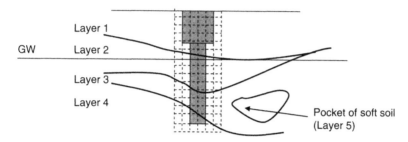

Figure 20.2   Boundary elements

- Only the elements at the soil pile boundary are represented.

- In this method, a full soil profile is not represented. As one can see, the isolated soft soil pocket is not represented.

- On the other hand, fewer elements would make the computational procedure much simpler than the finite element method.

20.3   Lateral Loading Analysis—Computer Software

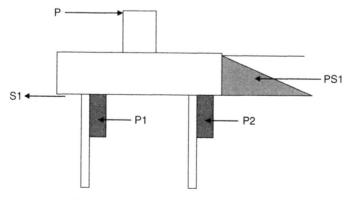

Figure 20.3   Lateral loading analysis

When a lateral load (P) is applied as shown, the following resistances are developed.

PS1 = passive soil resistance due to pile cap on one side of the pile cap
S1 = skin friction at the base of the pile cap
P1 and P2 = lateral soil resistance of piles.

If the pile cap is connected to other pile caps with tie beams, there will be resistance due to tie beams as well.

Computer Programs for Lateral Load Analysis

COM624P: This program is capable of analyzing laterally loaded single piles. It is based on the finite difference technique.

FLPIER (Florida Pier): This program is specially designed to analyze bridge pier systems. It is capable of analyzing pile groups/pile cap systems. FLPIER is based on the finite element method.

Lateral Loading Analysis Using Computer Programs

Input parameters to computer programs are twofold.

1. Pile parameters
2. Soil parameters

Pile Parameters

- Pile diameter
- Center-to-center spacing of piles in the group
- Number of piles in the group
- Pile cap dimensions

Soil Parameters for Sandy Soils

Soil parameters should be provided to the computer for each stratum.

- Strata thickness
- ϕ' value of the strata
- Coefficient of subgrade reaction (k)

Note: ϕ' value of sandy soil can be calculated using the following equation.

$$\phi' = 53.881 - 27.6034 \cdot e^{-0.0147N} \quad \textit{(Peck et al., 1974)}$$
(N = Average SPT value of the strata)

Note: The coefficient of subgrade reaction (k) can be obtained using the following table.

Coefficient of Subgrade Reaction (k) vs. N (SPT)

SPT (N)	8	10	15	20	30
k (kN/m³)	2.67 E-6	4.08 E-6	7.38 E-6	9.74 E-6	1.45 E-6

Source: Johnson & Kavanaugh (1968).

Similarly, soil parameters for other strata also need to be provided.

Soil Parameters for Clayey Soils

Soil parameters required for clayey soils:

- S_u (undrained shear strength. S_u is obtained by conducting unconfined compressive strength tests.)

- ε_c (strain corresponding to 50% of the ultimate stress. If the ultimate stress is 3 tsi, then ε_c is the strain at 1.5 tsi.)

- k_s (coefficient of subgrade reaction)
  The coefficient of subgrade reaction for clay soils is obtained from the following table.

Coefficient of Subgrade Reaction vs. Undrained Shear Strength

	Average Undrained Shear Strength (tsf)		
	(0.5–1) tsf	(1–2) tsf	(2–4) tsf
k_s (static) lbs/in3	500	1,000	2,000
k_s (cyclic) lbs/in3	200	400	800

Source: Reese (1975).

References

Johnson, S.M., and Kavanaugh, T.C., *The Design of Foundations for Buildings*, McGraw-Hill, New York, 1968.

Peck R., et al. *Foundation Engineering*, John Wiley and Sons, New York, 1974.

Reese, L.C., "Field Testing and Analysis of Laterally Loaded Piles in Stiff Clay," *Proceedings—Offshore Technology Conference*, Vol. II, Houston, TX, 1975.

Freeware from the Federal Highway Authority

The Federal Highway Authority (FHWA) provides computer software for geotechnical engineering professionals. The following Website link can be used to download the free programs.

http://www.fhwa.dot.gov/engineering/geotech/software/softwaredetail.cfm#driven

For pile design work, the FHWA provides the following programs.

20.4   Spile

Spile is a versatile program that determines the ultimate vertical static pile capacity. The program is capable of computing the vertical static pile capacity in clayey soils and sandy soils.

Pile Group Analysis: Pile groups need to be able to support vertical loads, lateral loads, bending moments, and uplift forces. In some cases, some of the piles may get damaged, and new piles need to be driven at a slightly different location. When the location of a pile is changed slightly, a new set of computations needs to be carried out.

Pile group program by Kalny Software company provides the following

a) Forces on individual piles in a pile group

b) Forces on corner piles for a given load.

21

Dynamic Analysis

Dynamic analysis is another technique used to evaluate the capacity of a pile, though presently many engineers prefer to use static analysis. Dynamic analysis is based on pile-driving data. Many dynamic equations are available, but their accuracy has been debated. The most popular formulas are:

- Engineering News Formula

- Danish Formula

- Hiley's Formula (more complicated than the other two formulas).

Dynamic formulas are better suited for cohesionless soils. The following comments were made by Peck, Hanson and Thornburn (1974).

All dynamic analysis formulas are unsound because their neglect of the time dependant aspects of the dynamic phenomena. Hence, except where well supported empirical correlations under a given set of physical and geological conditions are available, the use of formulas apparently superior to the Engineering News Formula is not justifiable.

In addition to neglecting the time-dependent aspect, pile dynamic formulas also ignore soil parameters and pile type.

21.1 Engineering News Formula

The energy generated by the hammer during free fall $= W_h \times H$

W_h = weight of the hammer; H = height of fall of the hammer (ft)

The energy absorbed by the pile $= Q_u \times S$

Q_u = ultimate pile capacity
 S = penetration in feet during the last blow to the pile. (Normally, average penetration during the last five blows is taken.)

Note: The work energy is defined as (force × distance).

During pile driving, the pile has to penetrate by overcoming a force of Q_u (ultimate pile capacity). Hence, the work energy is given by $Q_u \times S$.

$$W_h \times H = Q_u \times S$$

There is elastic compression in the pile and the pile cap. These inefficiencies are represented by a constant (C).

Hence, $W_h \times H = Q_u \times (S + C)$; C is an empirical constant.
C = 1 in. (0.083 ft) for drop hammers and C = 0.1 in. (0.0083 ft) for air hammers.

$$\text{Engineering News Formula} \rightarrow Q_u = \frac{W_h \times H}{(S + C)}$$

- Units of H, S, and C should be the same.
  Design Example: A timber pile was driven by an air hammer weighing 2 tons with a free fall of 3 ft. The average penetration (S) of the last five blows was found to be 0.2 ins. Use the Engineering News Formula to obtain the ultimate pile capacity (Q_u).

$$\text{Engineering News Formula} \rightarrow Q_u = \frac{W_h \times H}{(S + C)}$$

Convert all units to tons and inches. The C value for steam hammers is 0.1 in.
 W_h = 2 tons; H = 3 x 12 in.; S = 0.2 in.; C = 0.1 in.;

$$Q_u = \frac{2 \times (3 \times 12)}{(0.2 + 0.1)} = 240 \, \text{tons} \, (2{,}135 \, \text{kN})$$

$$Q(\text{allowable}) = 240/6 = 40 \, \text{tons}; \, (355 \, \text{kN})$$

Normally, a FOS of 6 is used.

21.2   Danish Formula

$$Q_u = \eta \frac{W_h \times H}{(S + 0.5\,S_0)} \quad \eta = \text{efficiency of the hammer}$$

Not all the energy is transferred from the hammer to the pile. A certain amount of energy is lost during the impact. η is used to represent the energy loss.

$$S_0 = \text{elastic compression of the pile} = \frac{2W_h \times H \times L}{A \times E}$$

L = length of the pile; A = cross-sectional area of the pile; E = modulus of elasticity of the pile.

21.2.1   Driving Criteria (International Building Code)

IBC states that the allowable compressive load on any pile should not be more than 40 tons, if driving formulas were used to compute the allowable pile capacity. According to IBC, if an engineer specifies an allowable pile capacity of more than 40 tons, wave equation analysis and pile load tests should be conducted.

Many building codes around the world depend on dynamic formulas for guidance.

The following table, based on the Engineering News Formula, has been adopted by the NYC Building Code.

Reference

Peck, R., Hanson, W., and Thornburn, T. "Foundation Engineering," 1974.
   IBC—International Building Code, 2005.

PART 4
Construction Methods

22

Pile Hammers

22.1 Introduction

- Driving a stick into the ground can be done by one person with a hammer.
- In earlier times, piles were driven by dropping a weight on the pile.
- The weight was lifted using various lever and pulley mechanisms.
- Guiding systems were developed to guide the weight so that it would fall vertically.

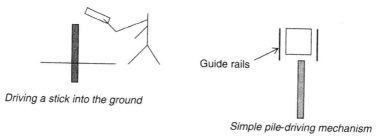

Driving a stick into the ground

Guide rails

Simple pile-driving mechanism

Figure 22.1   Pile driving

In the simple pile-driving mechanism shown in Figure 22.1, the following components are used.

1. A weight
2. A lifting mechanism (in this case a pulley)
3. Guide rails to make sure that the hammer falls vertically.

Early engineers who used this simple mechanism found out that it is more productive to use a smaller fall height. If the fall height is increased, then the time taken for a blow also will increase. It was more productive to have many low-energy blows than few high-energy blows. It was noted that low-energy blows could also minimize damage to piles.

22.2   Steam-Operated Pile Hammers

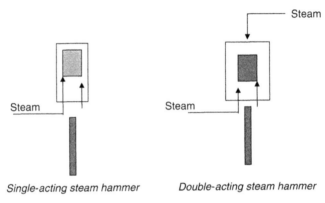

Single-acting steam hammer   Double-acting steam hammer

Figure 22.2   Steam operated pile hammers

22.2.1   Single-Acting Steam Hammers

Steam was allowed to enter the pile hammer chamber from the bottom. The hammer was lifted due to steam pressure. After lifting the hammer, the steam inflow is cut off. The hammer would then drop onto the pile.

22.2.2   Double-Acting Steam Hammers

Steam was allowed to enter the pile hammer chamber from the bottom. The hammer was lifted due to steam pressure. After lifting the hammer, the steam inflow was cut off. When the hammer started to drop, a second stream of steam was sent to the chamber from the top. The hammer falls faster due to gravity and the steam pressure above.

Note: Today compressed air is used instead of steam.

22.3   Diesel Hammers

Mechanism

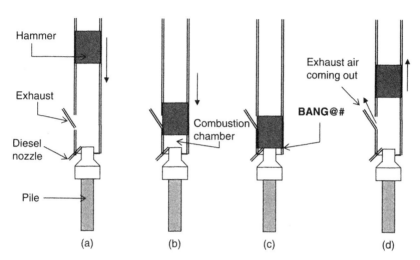

(a)   (b)   (c)   (d)

Figure 22.3   Diesel hammers. (a) The hammer is raised and ready to fall. (b) The hammer is dropping. The combustion chamber is filled with compressed air. At this point diesel is injected into the combustion chamber through the nozzle. (c) BANG! . . . . . Impact. When the impact occurs, diesel in the combustion chamber ignites and an explosion occurs. (d) The explosion inside the combustion chamber raises the hammer. Exhaust comes out of the exhaust air outlet. The cycle repeats.

22.3.1   Single-Acting Diesel Hammers

- The picture shown above is of a single-acting diesel hammer. Double-acting hammers have two combustion chambers. (See Figure 22.4.)

- Single-acting hammers have open-end tops, while double-acting hammers have closed-end tops. Diesel would be injected into both lower and upper chambers in double-acting hammers.

- In the case of single-acting diesel hammers, the hammer moves downward only because of gravitational force.

- In the case of double-acting diesel hammers, the hammer moves downward due to gravitational force and the explosion force in the *upper* combustion chamber.

- Double-acting hammers are capable of generating much more force than single-acting hammers.

22.3.2 Double-Acting Diesel Hammers

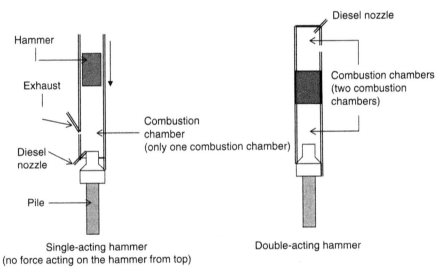

Figure 22.4   Single-acting and double-acting hammers

- Noise: Another major complaint against diesel hammers is the "noise." Special diesel models available from major manufacturers make less noise.

- Operation in cold weather: Both single- and double-acting diesel hammers are known to operate well under cold weather conditions.

- Hammer as a pile extractor: Single-acting hammers cannot be used as a pile extractor. Double-acting hammers are inverted and can be used as a pile extractor.

- Frequency: Single-acting hammers have a frequency of 50 to 60 blows per minute, while double-acting hammers could be as high as 80 blows per minute.

- Soft soil conditions: Diesel hammers can stall when driving in soft soil conditions.

- Air pollution: It is not a secret that diesel hammers create diesel exhaust after each stroke. For this reason, many engineers are reluctant to specify diesel hammers for urban pile-driving work. The latest models have a much better track record for cleanliness. This aspect needs to be investigated prior to specifying a diesel hammer.

Diesel Hammer Manufacturers

Delmag (www.delmag.de), Berminghammer (www.Berminghammer. com), APE (www.apevibro.com), MKT (www.mktpileman.com), ICE (www.iceusa.com), HMC (www.hmc-us.com), Mitsubishi, Kobe.

- The energy of diesel hammers can be as high as 500,000 lb/ft. A large diesel hammer D200-42 by Delmag imparts energy of 500,000 lb/ft to the pile at maximum rating. It has a hammer of 44,000 lbs (22 tons) and a stroke of 11 ft. Furthermore, it can provide 30 to 50 blows per minute as well.

- On a smaller side of the scale, D2 by Delmag has a hammer of 484 lbs with a stroke of 3 ft 8 in. providing an energy of 1,000 lb/ft with 60 to 70 blows per minute.

22.3.3   Environmental Friendly Diesel Hammers

- As an answer to the air pollution problem, a new generation of diesel hammers is being manufactured that uses Bio Diesel. Bio Diesel is made of soybean oil and is both nontoxic and biodegradable.

- Bio Diesel hammers can be used for projects in urban areas where exhaust gases are regulated.

- Example: Bio Diesel hammer by ICE (www.iceusa.com).

22.4   Hydraulic Hammers

- Instead of air or steam, hydraulic fluid is used to move the hammer.
- Hydraulic hammers are much cleaner because they do not emit exhaust gases.

- Hydraulic hammers are less noisy than most other impact hammers.

- Unfortunately, hydraulic hammers are expensive to buy and rent.

- Hydraulic hammers are capable of working underwater as well.

- Unlike steam or air hammers, the energy of hydraulic hammers can be controlled.

Mechanism

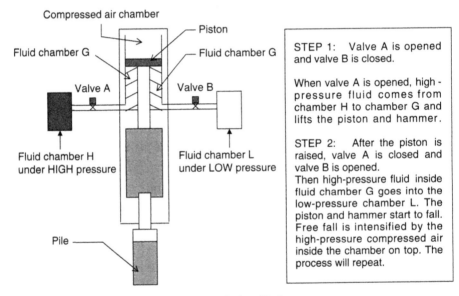

Figure 22.5 Hydraulic hammer

- Any free-falling object usually falls under gravitational acceleration g. But due to high-pressure compressed air on the top chamber, acceleration of the hammer can be raised to 2 g or more.

- Large hydraulic hammers can impart energy as high as 2.5 million lb/ft on the pile.

- For example, MHU 3000T by MENCK can impart energy of 2.43 million lb/ft. It has a striking part weighing 198.4 tons with a stroke of 5.5 ft.

Energy = Weight of striking part × Stroke = 396, 800 lbs × 5.5 ft
= 2.28 million lb/ft.

This giant is also capable of providing 40 blows per minute.

(Additional energy is obtained from the high-pressure gas inside the chamber above the piston.)

- S-2300 hammer by IHC has an energy rating of 1.6 million lb/ft with a hammer weighing 115 tons. It can provide 30 to 80 blows per minute as well.

  Stroke of S-2300 hammer (assuming free fall) = 1,600,000/(115 × 2000) = 6.95 ft

  Stroke = Energy/Weight of the striking part of the hammer

  But the stroke of the S-2300 hammer is only 3.5 ft. Higher energy is obtained from the high-pressure gas inside the chamber above the piston.

- IHC provides two hammer types. The S series by IHC has a lighter hammer and high hammer speed; these hammers are ideal for steel piles. IHC SC series hammers have heavier hammers and slow hammer speed; these are good for concrete piles.

- On the smaller side of the scale, HMC 28H has an energy rating of 21,000 lb/ft with a hammer of 16,500 lbs with a stroke of 1 ft.

- Energy can be varied: Energy imparted to the pile can be controlled by varying the pressure in the gas chamber above the piston. The control panel can be hooked to a printer to obtain printouts.

- Control panel: Electronic control panels are available with latest models. The typical control panel provides energy, engine rpm, and so on.

- Noise reduction: Special housings have been developed for the purpose of noise reduction.

- Energy per blow: The operator can adjust the energy per blow.

- Blow rate: Hydraulic hammers are capable of providing high blow rates 50 to 200 blows per minute.

22.5   Vibratory Hammers

Mechanism

Vibratory hammers consist of three main components.

1. Gear case

2. Vibration suppressor

3. Clamp

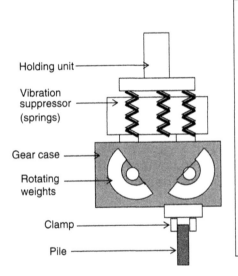

- The gear case has eccentric rotating weights.
- Due to the eccentricity of the weight, the gear case moves up and down (or vibrates up and down).
- These vibrations are transferred to the pile below.
- The suppressor above the gear case suppresses any up-and-down vibrations. The suppressor is made up of springs.
- Due to the suppressor, the holding unit will not feel any vibrations.
- The holding unit can be attached to an arm of a backhoe, hung by a crane or by a helicopter in rare occasions.
- Up-and-down movement of the gear case is dependent on the RPM (revolutions per minute) of weights.

Holding unit

Vibration suppressor (springs)

Gear case

Rotating weights

Clamp

Pile

Figure 22.6   Vibratory hammer

Principle of the Vibratory Hammer

- Imagine a weight attached to a rod. If the weight is attached to the rod at the center of gravity, there will not be any vibrations during rotation.

- If, however, the weight is attached from a point other than the center of gravity, then vibrations will occur. (See Figure 22.7.)

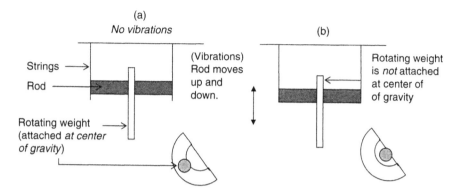

Figure 22.7   Vibratory hammer mechanism

- In the case shown in Figure 22.7a, the rod will not move up and down when the mass is rotated. Thus, the rotating mass is attached to the rod at the center of gravity.

- In the case shown in Figure 22.7b, the rod will *move up and down* when the mass is rotated. In this case, the rotating mass is *not* attached to the rod at the center of gravity.

- This principle is used for vibratory hammers. The gear case contains a rotating mass attached to a rod as shown in Figure 22.7b, with an eccentricity.

- Eccentricity = distance between point of attachment and center of gravity

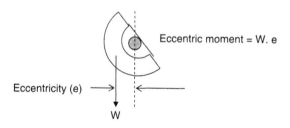

Figure 22.8   Eccentric moment

Amplitude: Up-and-down movement of the vibratory hammer. Amplitude could be 0.2 to 1.5 in.

Frequency: The number of up-and-down movements of the vibratory hammer per minute. Typically, frequency ranges from 1,000 to 2,000 rpm (revolutions of rotating weights per minute).

Power Pack: Vibratory hammers are usually powered by a power pack.

Some Vibratory Hammer Specifications

Manufacturer	Model	Frequency (RPM)	Amplitude (in)	Max. Pull (Tons)	Pile Clamp Force (Tons)	Eccentric Moment (in. lbs)
DAWSON	EMV 300	2,400	0.58	8.8	40	400
TRAMAC	428B	3,000	0.35	14	56	348
HPSI	40E	2,200	0.875	15	30	400
ICE	216E	1,600	1.02	45	50	1,100

Electronic monitoring: Most vibratory hammers come with electronic monitoring devices. These devices provide information such as frequency, amplitude, maximum pull, and eccentric moment at any given moment.

Sands and clays: Vibratory hammers are ideal for sandy soils. When the pile vibrates, the soil particles immediately next to the tip of the pile liquefy and resistance to driving diminishes. In clay soils, this process does not occur. Vibratory pile hammers are not usually effective for driving in clay soils. If one encounters clay soil during driving, larger amplitude should be used for clay soils. This way, clay soil can be moved and pile can progress. When a smaller amplitude is selected, pile and surrounding clay can move up and down together without any progress.

Popular dynamic formulas cannot be used: One major disadvantage of vibratory hammers is their inability to use popular dynamic formulas. (Many engineers and city codes are used to specifying the end of pile driving using a certain number of blows per foot.)

22.5.1 Resonance-Free Vibratory Pile Drivers

- Highest amplitude occurs at the startup and finish of a vibratory pile driver. At startup, weights inside the vibratory driver accelerate to achieve a high velocity. This induces a high amplitude at the beginning. High amplitude can cause damage to nearby buildings.

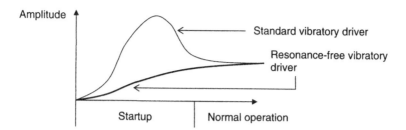

Figure 22.9 Amplitude in a vibratory hammer

- At startup, standard vibratory drivers generate high amplitudes as shown in Figure 22.9. Specially designed resonance-free vibratory drivers do not generate high amplitudes during startup.

- Resonance-free vibratory drivers may be selected when driving near buildings.

22.6   Pile-Driving Procedure

The following procedure is generally used during the driving of piles.

- A stake is driven at the location of the pile.

- The pile is straightened and kept upright on the location marked.

- The plumbness of the pile is checked.

- The pile-driving hammer is lowered to the top of the pile.

- Few light blows are given and checked for plumbness.

- Full driving starts.

- The rate of penetration is recorded.

Rate of penetration = Number of blows per inch

- The pile is driven to the planned depth. In some cases, pile driving is stopped when the required rate of penetration is achieved.

- The inspector should keep an eye on rate of penetration of the pile. Unusually high blows per inch may indicate boulders or bedrock. Some piles (especially timber piles) get damaged if they hit a boulder.

- After the pile is driven as per the required criteria, the pile is cut off from the top.

23

Pile Inspection

23.1 Pile-Driving Inspector's Checklist

- Review of the Geotechnical Engineering Report
- Inspection of piles prior to installation
- Inspection of pile-driving equipment (prior to driving and during driving)
- Inspection of the pile-driving procedure
- Maintenance of a driving record log

23.2 Review of the Geotechnical Engineering Report

The inspector should read and understand the Geotechnical Engineering Report. He should be aware of soil conditions, expected capacity of piles, specifications of piles and driving equipment, and pile-driving criteria.

23.3   Inspection of Piles Prior to Installation

• Inspection of Timber Piles: The piles should comply with the specifications of the project, and the timber should be of the species specified. Tip and butt diameter should be measured and recorded for each pile. These diameters should not vary significantly from the specified diameters. Timber piles should be straight. Timber piles that are not as straight as specified should be rejected. The inspector should look for any damage due to fungus attack. If steel shoes are specified, these shoes should be properly attached.

• Inspection of Steel H-piles: The inspector should check for the steel grade, section dimensions, and length. Pile shoes and pile-splicing techniques should be in accordance with the specifications. Piles should be free of damage and corrosion.

• Inspection of Concrete Piles: Check for the dimensions, slump report, and strength report of concrete. Specifications normally require a minimum waiting period after casting, prior to driving. It is common for concrete piles to have slight damage during transportation. The inspector should check for this damage and if it is significant, that pile should be rejected.

23.4   Inspection of Pile-Driving Equipment (prior to driving and during driving)

• The pile-driving rig should be in accordance with specifications.

• The pile hammer should be as specified. A higher energy rating than specified energy can damage the pile, while a smaller energy rating may not produce the driving energy required.

• The helmet of the pile should properly fit the pile. If the helmet does not fit the pile properly, that could cause damage to the pile.

• Determine the weight of the ram (for drop hammers and air/steam hammers).

23.5   Pile-Driving Inspection Report

Pile Information	The Equipment	The Hammer
• Pile type: wood__H pile__concrete__pipe pile__ • Pile diameter: (Wood) Tip_____But_____ • H pile: Section _____ Weight per foot _____ • Pipe Pile: Outer diameter _____ Inner diameter _____	Rig Type:_____ Mandrel:_____ Follower_____ Cap: Type____Wt:_____	Make:_____ Model:_____ Rated Energy:_____ Wt of Ram:_____ Stroke:_____ Hammer cushion: Thickness__

Project Name:_____ Address_____ Project #_____

Ground Elevation_____ Pile Toe Elevation_____
Length Driven_____
Pile Cutoff Elevation_____ Pile Splicing Information_____

Contractor Information:
Contractor's Name:_____ Field Foreman's Name:_____
Contractor's Address:_____

Pile Inspector's Name_____ Pile Inspection Company:_____
Weather: Temp:_____ Rain_____

Ft	No. of Blows	Speed (blows per minute)	Remarks	Ft	No. of Blows	Speed (blows per minute)	Remarks
1 2 3 4 25					26 27 28 29 50		

23.6 General Guidelines for Selecting a Pile Hammer

23.6.1 Single-Acting Steam and Air Hammers

- Dense sands and stiff clays need heavy hammers with low blow counts. This makes single-acting hammers ideal for such situations.

23.6.2 Double-Acting Steam and Air Hammers

- Double-acting hammers have light hammers compared to single-acting hammers with the same energy level. Light hammers with high-velocity blows are ideal for medium-dense sands and soft clays.

23.6.3 Vibratory Hammers

- Vibratory hammers for concrete and timber piles should be avoided. Vibratory hammers can create cracks in concrete.

- Vibratory hammers for clayey soils should also be avoided. Vibratory hammers are best suited for loose to medium sands.

- Vibratory hammers are widely used for sheetpiles since it may be necessary to extract and re-install piles. Piles can be readily extracted with vibratory hammers.

- In loose to medium soil conditions, sheetpiles can be installed at a much faster rate by vibratory hammers.

23.6.4 Hydraulic Hammers

- Hydraulic hammers provide an environmentally friendly operation. Unfortunately, the rental cost for these hammers is high.

23.6.5 Noise Level

Another important aspect of selection of piles is the noise level.

Installation Method	Observed Noise Level	Distance of Observation
Pressing (jacking)	61 dB	7 m
Vibratory (medium freq.)	90 dB	1 m
Drop hammer	98–107 dB	7 m
Light diesel hammer	97 dB	18 m

Source: White et al. (1999).

Reference

White, D.J., et al. *Press in Piling: The Influence of Plugging on Driveability*, 8th International Conference on Plugging on Driveability, 1999.

23.7   Pile Driving Through Obstructions

- Most pile-driving projects encounter obstructions. The types of obstructions usually encountered are

  1. Boulders

  2. Fill material (concrete, wood, debris)

23.7.1   Obstructions Occurring at Shallow Depths

If the obstruction to pile driving is occurring at a shallow depth, a number of alternatives are available.

Excavate Obstructions Using a Backhoe

Excavate and remove obstructions   Backfill with clean fill

Figure 23.1   Obstructions

This method may be cost prohibitive for large piling projects. The reach of most backhoes is limited to 10 to 15 ft and any obstruction below that level would remain.

Use Specially Manufactured Piles

Most pile-driving contractors have specially designed thick-walled pipe piles or H-piles that penetrate most obstructions. Usually, these piles are driven to penetrate obstructions. After penetrating the obstruction, a special pile is removed and the permanent pile is driven. The annulus is later backfilled with sand or grout.

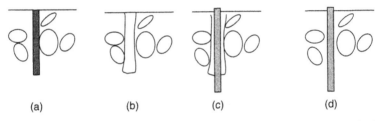

(a)   (b)   (c)   (d)

Figure 23.2 Driving through obstructions. (a) Drive the stronger pile (special pile) through obstructions. (b) Remove the special pile. (The hole is shown here). (c) Insert the permanent pile and drive it to the desired depth. (d) Grout the annulus or fill it with sand.

Obstructions Occurring at Any Depth

The second method above can be implemented for deeper obstructions as well, provided the hole remains open. Unfortunately, driving a special pile deep is somewhat equivalent to driving two piles. Other methods exist to tackle obstructions at deeper levels.

Spudding

Spudding is the process of lifting and dropping the pile constantly until the obstruction is broken into pieces. Obviously, spudding cannot be done with lighter piles (timber or pipe piles). Concrete piles and steel H-piles are good candidates for spudding.

Spudding generates extremely high stresses. The spudding piles should be approved by the engineer prior to use in the site since spudding could damage the pile.

Augering and Drilling

It is possible to auger through some obstructions. Usually, local contractors with experience in the area should be consulted prior to specifying augering or drilling through obstructions.

Toe Strengthening of Piles

The toe of piles can be strengthened using metal shoes to penetrate obstructions.

23.8   Pile Hammer Selection Guide

In most cases piles are selected using wave equation programs. If wave equation analysis is not conducted, the following table can be used as an approximate guide. (The table was prepared by adapting the table presented in "Pile Driving Equipment," U.S. Army Corps of Engineers, July 1997).

23.8.1 Sandy Soils

SPT (N) Value	Soil Type	Timber Piles	Open-End Pipe Piles	Closed-End Pipe Piles	H-Piles	Sheetpiles	Concrete Piles
0–3	Very loose	DA, SA (A, S, H)	DA, SA, V (A, S, H)	(A, S, H)	DA, SA, V (A, S, H)	DA, SA, V (A, S, H)	DA, SA (A, S, H)
4–10	Loose	DA, SA (A, S, H)	DA, SA, V (A, S, H)	DA, SA, V (A, S, H)	DA, SA, V (A, S, H)	DA, SA, V (A, S, H)	DA, SA (A, S, H)
10–30	Medium	SA (A, S, H)	DA, SA, V (A, S, H)	DA, SA, V (A, S, H)	DA, SA, V (A, S, H)	DA, SA, V (A, S, H)	SA (A, S, H)
30–50	Dense	SA (A, S, H)	DA, SA, V (A, S, H)	SA, V (A, S, H)	SA, V (A, S, H)	DA, SA, V (A, S, H)	SA (A, S, H)
50+	Very dense	SA (A, S, H)	SA (A, S, H)	SA (A, S, H)	SA (A, S, H)	DA, SA, V (A, S, H)	SA (A, S, H)

Legend: DA = Double Acting; SA = Single Acting; A = Air, S = Steam, H = Hydraulic, V = Vibratory

23.8.2 Clay Soils

SPT (N) Value	Soil Type	Timber Piles	Open-End Pipe Piles	Closed-End Pipe Piles	H-Piles	Sheetpiles	Concrete Piles
0–4	Very Soft	DA, SA (A, S, H)	DA, SA, V (A, S, H)	DA, SA (A, S, H)	DA, SA, V (A, S, H)	DA, SA, V (A, S, H)	DA, SA (A, S, H)
4–8	Medium	DA, SA (A, S, H)	DA, SA, V (A, S, H)	SA (A, S, H)	DA, V (A, S, H)	DA, SA, V (A, S, H)	SA (A, S, H)
8–15	Stiff	SA (A, S, H)	DA, SA (A, S, H)	SA (A, S, H)	DA, SA (A, S, H)	DA, SA (A, S, H)	SA (A, S, H)
15–30	Very stiff	SA (A, S, H)	SA (A, S, H)	SA (A, S, H)	SA (A, S, H)	SA (A, S, H)	SA (A, S, H)
30+	Hard	SA (A, S, H)	SA (A, S, H)	SA (A, S, H)	SA (A, S, H)	SA (A, S, H)	SA (A, S, H)

23.9   Pile Heave and Re-Driving

- When driving piles in a group, nearby piles heave due to driving.

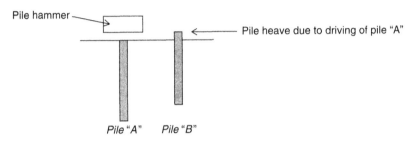

Figure 23.3   Pile heave

- It is important to re-drive piles that have been heaved by more than 1 in. (Local codes need to be consulted by engineers with regard to pile heave.)

23.9.1   Case Study (*Koutsoftas, 1982*)

In this case study, groups of piles were driven and pile heave was measured.

Site Conditions

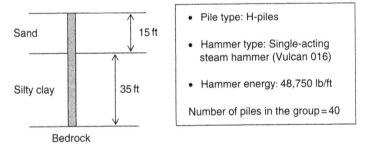

Figure 23.4   Pile – soil diagram

Pile group   = 5 × 8 cluster,
Pile type   = H-piles, HP 14 × 117, Grade 50 steel
Pile spacing = 3.5 ft center to center

Pile-driving sequence: One row at a time. Started from the southern-most row and moved north, driving row by row. (See Figure 23.5.)

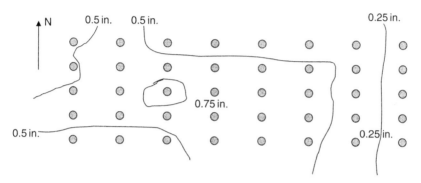

Figure 23.5   Pile location plan

Pile Heave Contours

The piles at the middle heaved more than the piles at the periphery.

Reference

Koutsoftas, D.C., "H-Pile Heave: A Field Test," *ASCE J. of Geotechnical Eng.* August 1982.

23.10   Soil Displacement During Pile Driving

Volume displacement during pile driving is significant in clayey soils. On the other hand, sandy soils tend to compact. Volume displacement of saturated clayey soils is close to 100%. In other words, if a 10-cu-ft pile is driven, 10 cu ft of saturated clay soil will need to find another home. Displaced soil usually shows up as surface heave or displace-ment of nearby sheetpiling or basement walls.

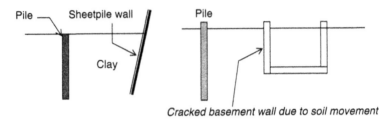

Figure 23.6   Possible problems due to pile driving

- Failure of a sheetpile wall due to pile driving is shown in Figure 23.6.

- The sheetpile wall probably would not have failed if the soil had been sand instead of clay.

- Pile driving could damage nearby buildings due to soil displacement or pore pressure increase. If there is a danger of affecting nearby buildings, pore pressure gauges should be installed.

- If pore pressure is increasing during pile driving, the engineer should direct the contractor to stop pile driving till pore pressure comes down to an acceptable level.

23.11   Pile Integrity Testing

- Low-strain methods (ASTM D 5882)

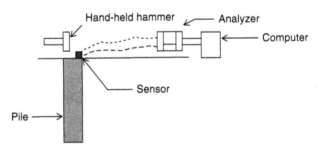

Figure 23.7   Pile driving analyzer

A hand-held hammer is used to provide an impact on the pile. The impact generates two types of waves: (1) longitudinal waves and (2) shear waves.

Waves generated due to the hammer impact travel along the pile and return. The sensor will relay the information to the analyzer. The analyzer makes computations and provides information regarding the integrity of the pile. If the pile is in good condition, both waves will return rapidly. If the pile has a broken section, that will delay the signal. Delay of each wave gives information such as necking, dog legging (bends), and cracked concrete. Depending on the product, some sensors are capable of recording the load on the pile, strain in the pile material, and stress relief.

• High-strain methods (ASTM D 4945)

Instead of a hand-held hammer, a pile-driving hammer also can be used. This technique provides the bearing capacity of soil in addition to pile integrity. The wave equation is used to analyze the strain and load on the pile.

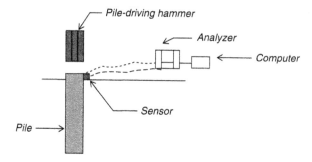

Figure 23.8   Pile driving analyzer with a pile hammer

Radar Analyzer

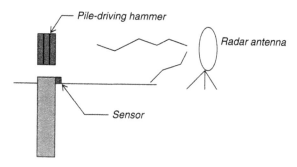

Figure 23.9   Radar analyzer

In this method, radar signals are sent to the pile hammer. Radar signals collide with the pile hammer and return to the antenna. From this information, the sensors in the antenna are able to detect the velocity of the pile. From the velocity, it is possible to compute the kinetic energy imparted to the pile. Some radar devices are capable of monitoring the stroke as well.

23.12   Use of Existing Piles

After demolition of a structure, piles of the old structure may remain. Cost of construction of the new structure can be reduced if old piles can be used. Many building codes allow use of old piles with proper procedures.

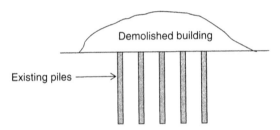

Figure 23.10   Use of existing piles

The design engineer should conduct a thorough survey of old piles. The following factors should be taken into consideration during the survey.

1. Capacity of old existing piles: If old Geotechnical Engineering Reports can be obtained, it may be possible to find pile-driving logs, pile load test data, and boring logs belonging to the old structure. These data can be used to find the capacity of old piles.

2. Reliability of old information: Piles may have been driven decades ago. The reliability of old information needs to be assessed.

3. Present status of old piles: Timber piles can be subject to deterioration. Steel piles may get rusted, and concrete piles are susceptible to chemical attack.

   • Test pits can be dug near randomly selected piles to observe the existing condition of piles

   • Few piles can be extruded to check their present status

4. Pile load tests: Pile load tests can be conducted to verify the capacity.

5. Pile integrity testing: Seismic wave techniques are available to check the integrity of existing piles. These techniques are much cheaper than pile load tests.

6. Approval from the building commissioner: Usually, the building commissioner needs to approve the use of old piles. The report of findings needs to be prepared and submitted to the local building department for approval.

23.13   Environmental Issues

23.13.1   Creation of Water-Migrating Pathways

• Piles are capable of contaminating drinking water aquifers. When piles are driven through contaminated soil into clean water aquifers, water migration pathways can be created. Water migrates from contaminated soil layers above to lower aquifers.

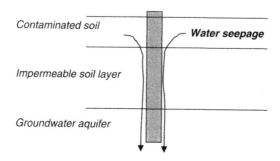

Figure 23.11   Water seepage along piles

As shown in Figure 23.11, water may seep along the pile into the clean water aquifer below.

- When a pile is driven or bored, a slight gap between soil and pile wall is created. Water can seep along this gap into clean water aquifers below. H-piles are more susceptible to creating water migration pathways than circular piles.

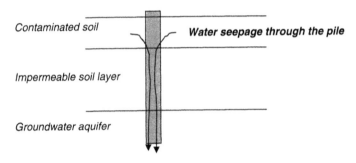

Figure 23.12 Wick effect

- Figure 23.12 shows water seeping *through* the pile into the lower soil strata. This situation is common to timber piles. Researchers have found that treated timber piles are less likely to transport water than untreated piles. Geotechnical engineers should be aware of this phenomenon when designing piles in contaminated soil conditions.

23.14 Utilities

Summary: Every geotechnical engineer should have good knowledge of utilities. Utilities are encountered during drilling, excavation, and construction. This chapter gives basic information regarding utilities.

General Outline of Utilities: In a perfectly planned city, utilities will be located as shown in Figure 23.13.

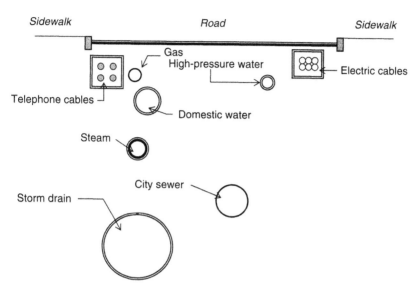

Figure 23.13   Typical utility distribution

Most cities were not built overnight; they were built over many centuries. Utility companies lay down utilities in accordance with the availability of space. Figure 23.13 shows how utilities would have been laid out, if there were no obstructions.

Electric, Cable TV, and Telephone Lines: Usually located at shallow depths closer to the edge of the road. These cables are enclosed in wooden or plastic boxes as shown in the figure.

Domestic Water: Domestic water lines are located relatively deep (approximately 4 to 6 ft). Domestic water lines consist of

• Mains

• Sub-mains

• Branch lines supplying buildings and houses

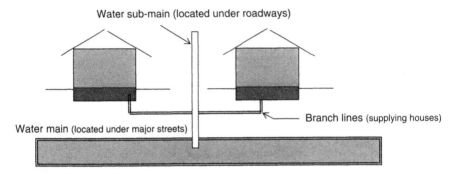

Figure 23.14   Water distribution lines

Note: The water main would go underneath a major street. Side roads would get water from this main. Sub-mains are connected to houses through branch lines. Main water lines are made of concrete or steel. Water lines are usually laid on top of stones and backfilled with clean sand. Prior to a water main, clean fill will be encountered. Water is carried through pipes by pressure.

High-pressure Water Lines: These lines are needed for fire hydrants. These lines are usually smaller in diameter and lie at shallow depths near the edge of roads.

Sewer Systems: Sewer lines are located at great depths away from water lines to avoid any contamination in the case of a sewer pipe failure. Usually, sewer lines are located 10 ft or more from the surface. Modern sewer lines are built of concrete. Old lines could be either clay or brick. Unlike water lines, sewer lines operate under gravity. Most sewer lines in major cities have been operating for many centuries without any interference from man; some believe that new creatures have evolved inside them.

Storm Drains: Storm drains are much larger than sewer pipes. They were usually built using concrete. (Older systems could be bricks or clay pipes.) Storm drains are responsible for removing water accumulated during storm events.

Rainwater is collected at catch basins and carried away by storm drains.

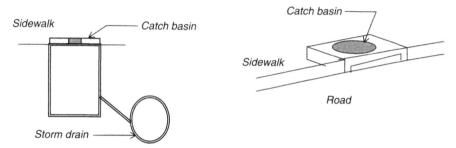

Figure 23.15   Storm drains

Combined Sewer: It is not unusual to connect sewer lines to storm drains. Since they are usually near discharge points, storm drains and sewer lines are combined. These lines are known as *combined sewer lines.*

24

Water Jetting

Water jetting is the process of providing a high-pressure water jet at the tip of the pile. Water jetting may ease the pile driving process by

- Loosening the soil
- Creating a lubricating effect on the side walls of the pile

24.1 Water Jet Types

Water jets can be external or internal. (See Figure 24.1.)

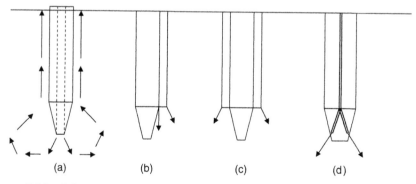

Figure 24.1 Water jet types. (a) Internal water jet pipe is shown here. Internal water jet pipes can be installed only in concrete piles and steel pipe piles. (b) External jet pipe attached to the side of the pile is shown. This is the usual procedure for timber piles. (c) Two jet pipes attached to opposite sides of the pile are shown here. (d) Internal jet pipe with two branches.

24.2   Ideal Water Pathway

Figure 24.1a shows the ideal water pathway. Water should return to the surface along the gap existing between the pile wall and soil. In some cases, water may stray and fill up nearby basements or resurface at a different location.

24.2.1   Pile Slanting Toward the Water Jet

• Piles tend to slant toward the water jet when driving.

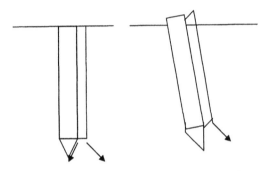

Figrue 24.2   Pile slanting due to water jet

• Pile slant toward the water jet can be avoided by providing two water jets symmetrically as shown in Figures 24.1c and 24.1d. Providing the water jet at the middle also (Figure 24.1a) is a solution to this problem.

Jet Pipe Movement

Many experienced pile drivers move the jet pipe up and down along the pile to ease driving. This process is specially conducted to keep the plumbness of the pile. This luxury is not available with internal jet pipes.

Water Jetting in Different Soil Types

Sand: Water jetting is ideal for dense sandy soils.
Silt: Water jetting can be successful in many silty soil beds. Yet, it is not unusual to have a clogged jet pipe. Clogging occurs when tiny

particles block the nozzle of the jet pipe. It is also important for the surrounding soil to close up on the pile after jetting is over. In some silty soils this may not happen. That would reduce the skin friction.

Clay: Water jetting can be used with different success levels in clay soils. Clay soils could clog the jet pipe as in the case of silty soils. Ideally, the water jets are supposed to travel back to the surface along the gap between pile wall and soil. In clay soils, this gap may be closed due to adhesion, and water may not be able to find a pathway to the surface. In such situations, jetting will not be possible.

Gravel: Water jetting is not recommended in loose gravel. Water travels through gravel (due to high permeability) and may not provide any loosening effect. Water jetting can be used in dense gravel with different levels of success.

Advantages of Water Jetting

- Water jetting is a silent process.

- Water jetting can minimize damage to piles.

24.3   Water Requirement

Case 1: Pile Driving in Dry Sand (water table is below the pile tip)

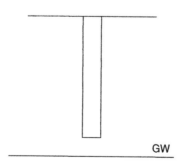

Figrue 24.3   Pile driving in dry sand

$$Q/D = 530\,(d_{50})^{1.3} \cdot L^{0.5} + 0 \cdot 1\,\pi \cdot L \cdot K$$

Q = flow rate of water required for pile jetting (units—cu m/hour)
D = pile diameter (m)

d_{50} = 50% passing sieve size
   L = pile length (m)
   K = permeability of soil (m/day)   (*Tsinker, 1988*)

Note: When the water table is located above the pile tip, the flow rate required will be less than the above computed value.

Case 2: Pile Driving in Dry Multilayer Sand (water table is below the pile tip)

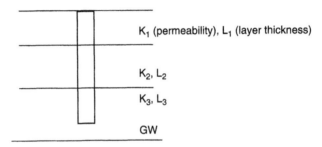

Figrue 24.4   Pile driving in multi sand layers

$$Q/D = 530\,(d_{50})^{1.3} \cdot L^{0.5} + 0.1\,\pi \cdot L \cdot K_m$$

$K_m = (K_1 \cdot L_1 + K_2 \cdot L_2 + K_3 \cdot L_3)/(L_1 + L_2 + L_3)$
K_m = combined permeability of sand layers

Determination of Required Pump Capacity

$$H = (Q^2 \cdot L_h)/C$$

   H = head loss in the hoses
   Q = flow rate (cu m/hour)
   L_h = total length of water supply hoses (m)
   C = head loss parameter (obtained from the table below)

For Rubber Hoses

Jet pipe internal diameter (mm)	33	50	65	76
C	50	200	850	2,000

Case 3: Pile Driving in Dry Sand (water table is above the pile tip)

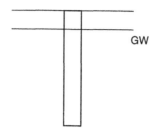

Figrue 24.5   Pile driving in sand (water table above the pile tip)

$$Q/D = 530\,(d_{50})^{1.3} \cdot L^{0.5} + 0.017\,\pi \cdot L \cdot K$$

Q   = flow rate of water required for pile jetting in cu m/hour
D   = pile diameter (m)
d_{50} = 50% passing sieve size
L   = pile length (m)
K   = permeability of soil (m/day)   (*Tsinker, 1988*)

Note: Saturated sand requires much less water than dry sand.

Reference

Tsinker, G.P., "Pile Jetting," *J of ASCE Geotechnical Eng.*, March 1988.

25

Cost Estimate for Pile-Driving Projects

As in any project, pile-driving projects consist of material and labor costs. The following guidelines help to generate a cost estimate for a piling project.

- Mobilization cost: Depends on the location, project size, and equipment.

- Material cost: Use the following chart.

Pile Type	Linear ft	Unit Cost	Total Cost
Wood piles, Class __ dia. ___ treated Y/N,			
Steel H Piles Size __			
Pipe Piles dia. ___, wall thickness ___			
Precast concrete piles Size __			
Concrete-filled pipe piles Size __			
Cast-in-situ concrete piles Size __			

- Pile shoes: No. of shoes required _____, Cost of shoes _____

- Pile-splicing cost: Material cost for splicing _____, Labor cost ____

- *Equipment Cost

  Driving rig make _____, Model _____
  Estimated driving rate _____, Total driving time _____,
  Allowance to move from pile to pile _____
  Rental rate _____, Cost of driving rig for the project _____
  Mandrel rate _____, Cost of mandrel for the project _____

- Labor

  Rig operator rate _____, Operator cost for the project _____
  Oiler rate _____, Oiler cost for the project _____
  Fireman rate _____, Fireman cost for the project _____
  Labor rate _____, Labor cost for the project _____

26

Pile Load Tests

26.1 Introduction

Pile load tests are conducted to assess the capacity of piles.

26.1.1 Theory

- Pile load test procedures can change slightly from region to region depending on local building codes.

- A pile is driven, and a load is applied to the pile.

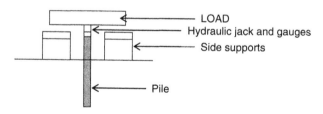

Figure 26.1 Pile load test

- Figure 26.1 is a highly simplified diagram of a pile load test. Nevertheless, it shows all the main parts that need to conduct a pile load test.

- Generally, to conduct a pile load test, one needs a driven pile, a load (usually steel and timber), a hydraulic jack, a deflection gauge, and a load indicator.

- The gauges should be placed to measure

    1. The load on the pile

    2. Settlement of the pile

Another Popular Configuration

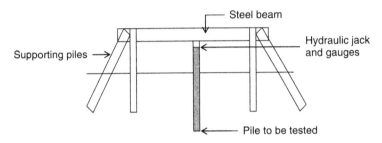

Figure 26.2   Pile load test using a steel beam

- In this configuration, a steel beam is held to the ground by supporting piles. A hydraulic jack is placed between the pile to be tested and the steel beam.

- The jack is expanded pushing the steel beam upward while pushing the test pile downward into the ground.

Pile Load Test Procedure

- Compute the ultimate pile capacity based on soil mechanics theory.

- Compute the design load using a suitable factor of safety

    Design Load (D) = Ultimate Pile Capacity (U)/FOS

- Generally, total test load is twice the design load. (This could change depending on local building codes.)

    Total Test Load (Q) = 2 × Design Load (D)

- Apply 12.5% of the test load and record the settlement of the pile every two hours.

- Readings should be taken until the settlement recorded is less than 0.001 ft during a period of two hours. When the pile settlement rate is less than 0.001 ft per two hours, add another 12.5% of the test load. Now the total load is 25% of the test load. The settlement is monitored as previously. When the settlement rate is less than 0.001 ft per two hours, the next load is added.

• The next load is 25% of the test load. Now the total load is 50%. The load is increased to 75% and 100% and settlement readings are taken.

• The load is removed in the same sequence, and the settlement readings are recorded. At least a one-hour time period should elapse during removal of loads.

• The final settlement reading should be recorded 48 hours after removal of the final load.

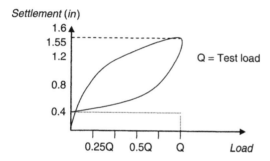

Figure 26.3   Load vs. settlement

• $Q = Maximum\ Test\ Load;$ Maximum test load (Q) is twice the design load. Settlement is marked along the Y-axis. The settlement increases during the application stage of the load and decreases during the removal stage of the load.

Gross settlement = the total settlement at test load (In this case it is 1.55 in.)
Gross settlement = settlement of pile into the soil + Pile shortening

• It is obvious that the pile compresses due to the load, and as a consequence the pile will shorten.

Net settlement = settlement at the end after the load is fully removed (In this case it is 0.4 in.)
Net settlement = settlement of pile into the soil after removal of the load

- When the load is released, the pile returns to its original length. In a strict sense, there may be a slight deformation even after the load is removed. In most cases, it is assumed that the pile comes to its original length when the load is removed.

- *The pile load test is considered to have failed if the settlement into the soil is greater than 1 in. at full test load*

  or

  if the settlement into the soil is greater than 0.5 in., at the end of the test after removal of the load.

26.2   Pile Load Test Data Form

Pile Information	The Equipment	The Hammer

• Pile type: wood: __ H pile: __
  concrete: __ pipe pile: __
• Pile diameter: (Wood)
  Tip: _____ Butt: _____
• H pile:
  Section: _____
  Weight per foot: _____
• Pipe pile:
  Outer diameter: _____
  Inner diameter: _____
	Rig Type: _____
	Mandrel: _____
	Follower: _____
	
	Cap:
	Type: __Wt: ____
		Make: _____
		Model: _____
		Rated energy: _____
		Wt of ram: _____
		Stroke: _____
		Hammer Cushion:
		Thickness: ___

Project Name: _____ Ground Elevation: _____ Pile Cutoff Elevation: _____	Address: _____ Pile Toe Elevation: _____	Project #: _____ Length Driven: _____
Pile Splice Data:	Depth of the Splice: _____	Type of the Splice: _____

Contractor Information:
Contractor's Name: _____
Contractor's Address: _____
Pile Inspector's Name: _____

Field Foreman's Name: _____

Pile Inspection Company: _____

Design Load:
Test Load:
Weather: Temp: _____ Rain: _____

Date	Time	Inspector's Name	Load (Tons)	Gauge Pressure (psi)	Settlement Gauge #1	Settlement Gauge #2	Remarks

27

Underpinning

27.1 Introduction

Underpinning is conducted mainly for two reasons.

1. To stop settlement of structures
2. To transfer the load to a lower hard stratum

27.1.1 Underpinning to Stop Settlement

Design Example 1. Underpin the strip footing shown in Figure 27.1 to stop further settlement.

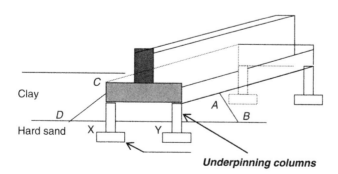

Figure 27.1 Underpinning with columns

27.1.2 Procedure

STEP 1: Excavate an area as shown in the figure (the excavation area is marked ABCD). Expose the area under the footing to erect underpinning columns. The structural integrity of the strip footing needs to be assessed prior to excavation of the footing. Normally 10 to 15% of a strip footing can be excavated in this manner.

STEP 2: Erect column footings X and Y as shown. Usually, columns are made of reinforced concrete.

Conduct the same procedure for the other end of the strip footing. Contractors usually buttress the excavated footing with timber, during construction of the concrete underpinning columns.

STEP 3: The following aspects need to be taken into consideration during the design process.

- Eventually, the total load will be transferred to underpinning piles.

- The central portion of the strip footing may crack due to sagging. If the strip footing is too long, underpinning columns may be necessary at the center of the strip footing as well.

27.2 Pier Underpinning

Scenario: A wall supported by a strip footing has developed cracks. These cracks have been attributed to uneven settlement in the strip footing. By conducting boreholes, a previously unnoticed pocket of soft clay was located (see Figure 27.2).

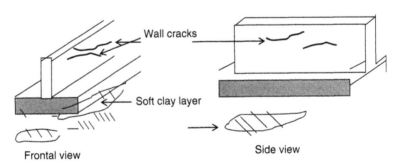

Figure 27.2 Structural damage due to weak soil

Solution: It is proposed to provide a concrete pier underneath the footing as shown below. The pier would extend below the clay layer to the hard soil layers below.

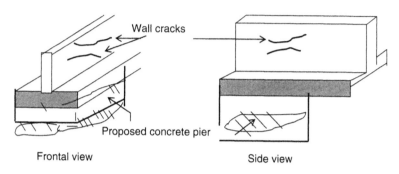

Frontal view   Side view

Figure 27.3   Pier underpinning

27.2.1   Pier Underpinning—Construction Procedure

Basically, a pit needs to be dug underneath the footing. In a strip footing, a pit can be dug underneath a small section.

STEP 1: Construct an approach pit.

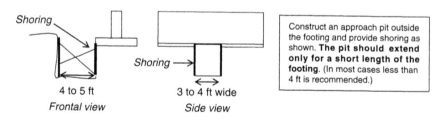

Figure 27.4   Approach pit

STEP 2: Generally, these approach pits are not more than 3 to 4 ft wide. In most cases, depth can be kept under 4 ft. The purpose of the approach pit is to provide space for the workers to dig underneath the footing and later concrete underneath the footing. All precautions should be taken to avoid any soil failure. Proper shoring should be provided to hold the soil from failing.

Excavate underneath the footing and provide shoring.

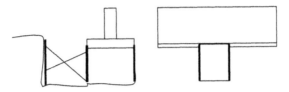

Figure 27.5   Excavate underneath the footing

STEP 3: Excavate to the stable ground.

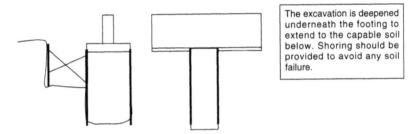

The excavation is deepened underneath the footing to extend to the capable soil below. Shoring should be provided to avoid any soil failure.

Figure 27.6   Excavate to the stable ground

STEP 4: Concrete within 2 to 3 in. from the footing.

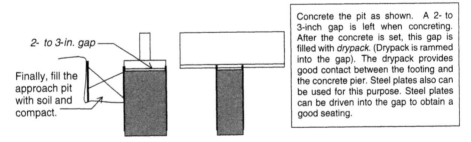

2- to 3-in. gap

Finally, fill the approach pit with soil and compact.

Concrete the pit as shown. A 2- to 3-inch gap is left when concreting. After the concrete is set, this gap is filled with *drypack*. (Drypack is rammed into the gap). The drypack provides good contact between the footing and the concrete pier. Steel plates also can be used for this purpose. Steel plates can be driven into the gap to obtain a good seating.

Figure 27.7   Concreting

After one section is concreted and underpinned, the approach pit for the next section is constructed and the procedure is repeated. It is important that each section not be more than 3 to 4 ft in length. Removal of soil under a large section of footing should be avoided.

27.3   Jack Underpinning

Pier underpinning is not feasible when the bearing strata are too deep. Manual excavation and construction of a pier will be too cumbersome. In such cases, jack underpinning can be a practical solution. A pit is constructed and a pipe pile is jacked underneath the footing. The load is transferred from the soil to the pipe pile.

Jack Underpinning of a Strip Footing

STEP 1: Construct an approach pit.

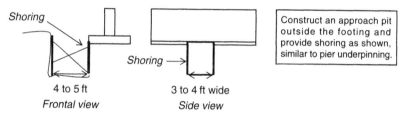

Figure 27.8   Construction of an approach pit

STEP 2: Excavate underneath the footing and provide shoring.

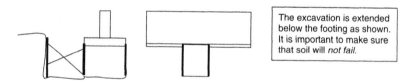

Figure 27.9   Excavate underneath the footing and provide shoring.

STEP 3: Set up the pipe pile and the jack.

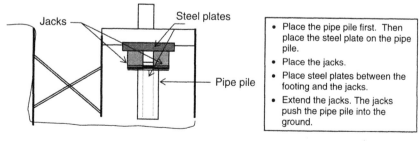

Figure 27.10   Setup the jack and the pipe pile

STEP 4: Push the pipe pile into the ground.

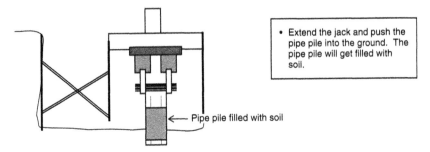

• Extend the jack and push the pipe pile into the ground. The pipe pile will get filled with soil.

← Pipe pile filled with soil

Figure 27.11 Push the pipe pile to stable ground

STEP 5: Remove the jacking assembly and clean the soil inside.

Remove the soil inside the pile.

• Remove the jacking assembly.
• Remove soil inside the pipe pile.
• Remove soil by hand augers, suction pumps, peel buckets, and water jets.
• Attach another piece of pipe and assemble the jacks.

Figure 27.12 Remove the jack and clean the soil

STEP 6: Attach another section of pipe and assemble the jacks.

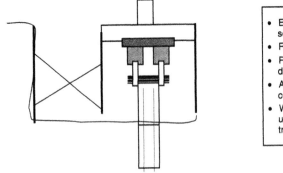

• Extend the jack and push the second pipe pile into the ground.
• Remove the soil inside the pile.
• Repeat the process until the desired depth is reached.
• After the desired depth is reached, concrete the pipe pile.
• Wait till the concrete is set and use drypack and steel plates to transfer the footing load to the pile.

Figure 27.13 Attach another section to the pile and repeat the procedure

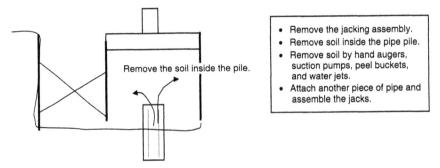

Notes: Instead of open-end pipe piles, H sections and Closed-end pipe piles also can be used. Closed-end pipe piles may be suitable in soft soil conditions. It may not be feasible to jack a closed-end pipe pile into a relatively hard soil layer. H sections may be easier to jack than a closed-end pipe pile. Extreme caution should be taken not to disturb the existing footing.

Monitoring Upward Movement of the Footing: When the pile has been jacked into the ground, the jacks will exert an upward force on the footing. The upward movement of the footing should be carefully monitored. In most cases, jacking the piles is conducted while keeping the upward movement of the footing at "zero." If the pile cannot be jacked into the ground without causing damage to the existing footing, a different type of a pile should be used. A smaller diameter pipe pile or smaller size H-pile should be selected in such situations.

27.4   Underpinning with Driven Piles

Strip footings can be underpinned with driven piles. An approach pit is constructed as before. A pile is driven outside the footing. The footing load is transferred to the pile through a bracket.

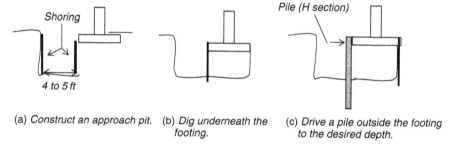

(a) *Construct an approach pit.*   (b) *Dig underneath the footing.*   (c) *Drive a pile outside the footing to the desired depth.*

Figure 27.14   Underpinning with driven piles

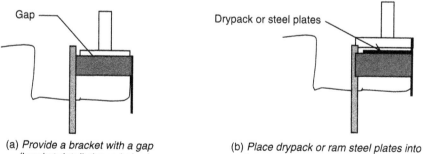

(a) *Provide a bracket with a gap (bracket detail shown below).*   (b) *Place drypack or ram steel plates into the gap.*

Figure 27.15   Provide a bracket

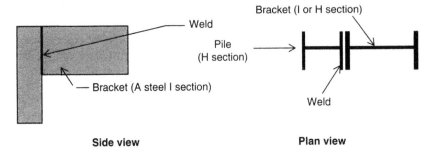

Figure 27.16   Bracket detail

27.5 Mudjacking (Underpinning Concrete Slabs)

Settling concrete slabs can be underpinned using a process known as *mudjacking*. In this technique, small holes are drilled, and a sand, cement, and limestone mixture is pumped underneath the slab to provide support.

Step 1: Drill holes in the concrete slab.
Step 2: Pump grout into the holes.

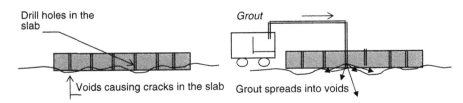

Figure 27.17   Mudjacking

The procedure is simple to understand. The engineer has to determine the following parameters:

Grout Pressure: Higher grout pressure can carry grout to greater distances. Voids under the slab can be filled at a faster rate with a high grout pressure. On the other hand, grout pressure has to be low enough, so that grout will not damage nearby existing structures.

Spacing between Holes: Greater spacing between holes requires high grout pressure. If the grout pressure has to be maintained at a lower level, spacing between the holes should be reduced. In general, holes are drilled at 3- to 4-ft spacing.

Composition of Grout: Grout for slab underpinning is mainly created using cement, sand, limestone, and water. Thicker grout can be used for coarser material, while a thinner grout needs to be used for fine sands. Silty soils and clays are difficult or impossible to grout.

Underpinning Companies

http://www.underpinning.com—*Underpinning & Foundation Construction*, New York-based company specializing in underpinning. (718)786-6557, Fax (718)786-6981.

http://www.haywardbaker.com—Nationwide contractor.

http://www.abchance.com—Nationwide contractor. Atlas helical pier technique. This technique is good for small-scale underpinning work. (573)682-8414.

http://www.casefoundation.com—Nationwide contractor.

http://www.mudjackers.com—Underpinning of concrete slabs using grout. (800)262-7584.

27.6   Underpinning: Case Study

• A 26-story building was settling on one side. The building was built on Frankie piles, which were constructed by hammering out a concrete base and concreting the top portion of the pile. (See Chapter 2.)

• The settlement took place rapidly.

• The building was 300 ft high, and the width was relatively small.

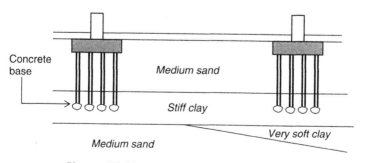

Figure 27.18   Underpinning case study

- A soil profile of the site is shown in Figure 27.18. During the boring program, a very soft clay layer occurring on the right-hand side of the figure was missed. For this reason, pile foundations on that side started to settle.

Pile Load Tests

- Pile load tests were conducted on the side that had the very soft clay layer. Pile load tests of single piles passed. This could be explained using a pressure bulb.

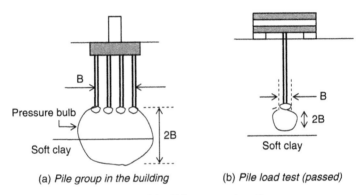

(a) *Pile group in the building*   (b) *Pile load test (passed)*

Figure 27.19   Pile group loading

- Typically, the pressure bulb of a footing extends 2B below the bottom of the footing (B = width of the footing). The width of the pile group is larger than the width of the concrete bulb of a single pile.

- In the case of the single pile (see Figure 27.19b), the pressure bulb does not extend to the soft clay layer below. This is not the situation with the pile group. The stress bulb due to the pile group extends to the soft clay layer. The stressed soft clay consolidated and settled. The success of pile load tests done on single piles could be explained using pressure bulbs.

The engineers were called upon to provide solutions to two problems: (1) Stop further settlement of the building; and (2) bring back the building to its original status. Further settlement was stopped by ground freezing.

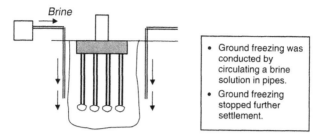

Figure 27.20   Ground freezing

Bringing the Building Back to the Original Position

- After further settlement was stopped by freezing the ground, an underpinning program was developed to bring the building back to the original level.

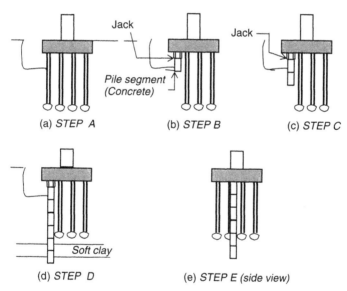

Figure 27.21   Underpinning procedure. (a) An approach pit was constructed as shown. (b) A concrete pile segment was placed, and a jack was assembled on top of the pile. (c) A concrete pile segment was jacked down using the pile cap above. The engineer had to make sure that the pile cap would be able to resist the load due to jacking. Usually, piles are jacked down one at a time. At any given time only one pile was jacked. (d) The pile segments are jacked down below the soft soil layer. (e) The new pile segments were jacked down between the existing piles without damaging the existing piles. Side view of a jacked down pile is shown in Figure 27.22.

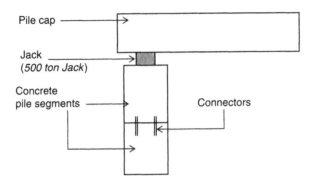

Figure 27.22   Underpinning with pile segments

Reference

Dummont, A., "The Underpinning of the 26 Story Compania Paulista De Seguros Building," Geotechnique 6, 1956.

28

Offshore Piling

- There are few fundamental differences between offshore piles and piles driven in land.

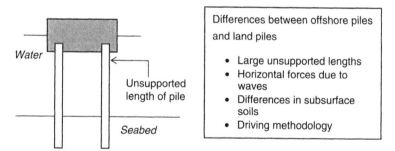

	Differences between offshore piles and land piles

- Large unsupported lengths
- Horizontal forces due to waves
- Differences in subsurface soils
- Driving methodology

Figure 28.1   Offshore piles

28.1   Seabed

- Almost all the offshore structures are constructed in continental crust.

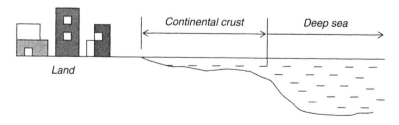

Figure 28.2   Continental crust and deep sea

- Continental crust was part of the land in the past. Land features such as canyons, eroded rocks, and land animal fossils could be seen in the continental crust. Deep-sea bed, however, had never been part of the land.

- Continental crust extends to hundreds of miles off Siberia while it stops a few hundred feet from the southern tip of South America. Extension of the continental crust varies from location to location.

- Most offshore structures are constructed in the continental crust. The major reason for this could be the shallowness of the water in the continental crust compared to the deep sea.

28.2 Soil Types in Continental Crust

Calcareous Sand: Calcareous sands are made of shells. Shells are living organisms. When the animal is dead, the shell is deposited and converted to calcareous sand.

   Calcareous sands create major problems for the pile designer. Calcareous sands do not develop much skin friction. On one occasion, a 180-ft-long pile driven in calcareous sand had a skin friction of almost zero! The pile was removed with a force almost equivalent to its self weight.

Dense Sands: Extremely dense sands are found in the continental shelf. Due to wave action, sand deposits can become very dense. In some occasions, ϕ, angle of soil, can be as high as 40 degrees.

Glacial Till: Glacial tills were formed by advancing glaciers and are extremely unpredictable since anything from boulders to clay particles can be encountered.

Silts: Highly overconsolidated silts can pose serious problems for pile driving. In such situations, water jetting is used to break up dense soils.

Clays: Clays in the continental shelf are overconsolidated due to water pressure.

28.3   Offshore Structures

A typical offshore structure would have three units.

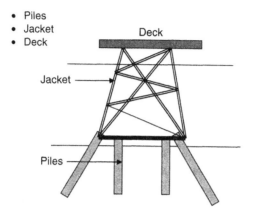

- Piles
- Jacket
- Deck

Figure 28.3   Offshore platform

28.3.1   Pile Installation

- *Driving with hammers*: This is the most popular installation method.

- *Drilling and grouting*: In this method, a hole is drilled and the pile is inserted. Then the annulus is grouted.

28.3.2   Pile Hammers

Single-Acting Air Hammers: Single-acting large air hammers have been more popular than other hammers. Air hammers with energy rating exceeding 1 million lb/ft are sometimes used to drive offshore piles. These hammers could weigh as much as 400 tons.

Diesel Hammers: Recently, Diesel hammers have also gained popularity. Instead of large air hammers, smaller Diesel hammers can be used with a higher number of blows per minute.

Hydraulic Hammers: Hydraulic hammers are capable of operating underwater and are gaining popularity.

Vibratory Hammers: Vibratory hammers may be useful in dense sands.

Water Jetting: Water jetting is increasingly used in offshore driving. Large pipe piles are manufactured with water jetting pipes in them.

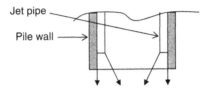

Figure 28.4   Water jetting

Water jets can have pressures up to 300 psi.

28.3.3   Drilling Prior to Driving

When boreholes indicate very dense soil conditions, a hole is drilled prior to driving the pile. The drill hole is made smaller than the pile in order to make sure that drilling does not affect the skin friction.

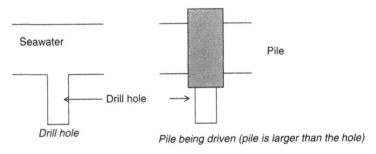

Figure 28.5   Bored piles

28.3.4   Pile Hammer Selection

In offshore pile driving, the following empirical rule is used to select hammers.

Rated energy of the hammer measured in ft kips

= 2 × Cross-sectional area of the pile wall measured in sq in

Using a hammer with higher energy can damage the pile. In some instances, pile-driving contractors may be forced to use larger hammers to reach the desired depth.

28.4   Drilled and Grouted Piles

Drilled and grouted piles are constructed by drilling a hole, then placing the pile and grouting. (See figures 28.6 and 28.7.)

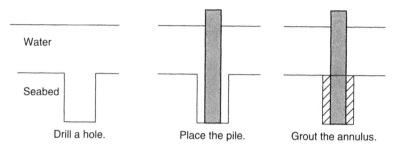

Figure 28.6   Drilled and grouted piles

• The drill hole should be at least 6 in. larger than the outside diameter of the pile. This method is used in situations where pile driving is difficult.

• In calcareous sands, this is the only method that could be used to generate skin friction. As mentioned earlier, calcareous sands do not provide much skin friction for driven piles. On the other hand, in drilled and grouted method, grout can migrate into voids to create a bond between grout and soil.

28.4.1   Belled Piers

Bells are used to increase the bearing capacity and uplift capacity of piles.

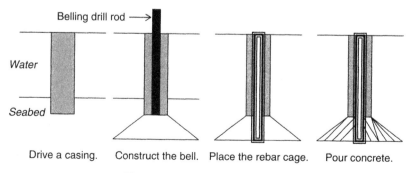

Figure 28.7   Belled caissons

- Placement of concrete below water needs special attention and skill. Usually, the Tremie method is used to concrete offshore piles.

References

Gerwick, B.C., Construction of Offshore Structures, John Wiley and Sons, New York, 1986.
McClelland, B., and Reifel, M.D., Planning and Design of Fixed Offshore Forms, Van Nostrand-Reinhold, New York, 1986.

29

Tie Beams, Grade Beams, and Pile Caps

The purpose of tie beams is to connect pile caps together. Tie beams will not carry any vertical loads such as walls etc.

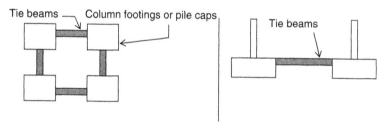

Figure 29.1 Tie beams

Unlike tie beams, grade beams carry walls and other loads.

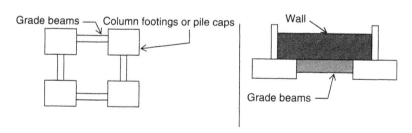

Figure 29.2 Grade beams

For this reason, grade beams are larger than tie beams.

After construction of pile caps or column footings, the next step is to construct tie beams and grade beams. Supervision of rebars needs to be conducted by qualified personnel prior to concreting.

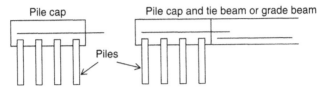

Figure 29.3   Pile cap and grade beams

Note: The contractor should provide rebars jutting out prior to concreting of pile caps.

This way when the grade beams are constructed, continuous rebars can be provided.

29.1 Pile Caps

The NYC Building Code recommends not using the vertical and lateral support capacities of pile caps. On the other hand, pile cap resting on a hard ground may be able to provide significant load-bearing capacity.

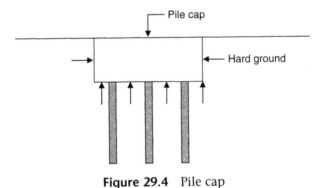

Figure 29.4   Pile cap

Vertical soil pressures acting on a pile cap are shown in Figure 29.4.

29.1.1 Sizing of the Pile Cap

STEP 1: The size of the pile cap depends on pile spacing and soil type.
STEP 2: Assume a pile type. In most cases, concrete auger cast piles are the cheapest and should be considered for feasibility. Closed-end pipe

piles are cheaper than any other driven pile. If there are obstructions (such as boulders or debris), then H-piles should be considered. Depending on site conditions, precast concrete piles, timber piles, or composite piles may be the most suitable pile type for the application.

STEP 3: Compute the ultimate bearing capacity of a single pile.

STEP 4: Obtain the design pile capacity of a single pile by applying a safety factor.

STEP 5: Find the number of piles required to carry the load.

STEP 6: Assume a center-to-center distance between piles. Typically, engineers use 2.5d to 3.0d. If center-to-center distance is too small, piles will damage each other. But, if the piles are too far, the pile cap will be too large and the cost will increase. If the site contains heavy obstructions, such as boulders, it is advisable to use a larger spacing.

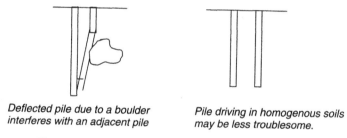

Deflected pile due to a boulder interferes with an adjacent pile

Pile driving in homogenous soils may be less troublesome.

Figure 29.5   Pile damage due to obstructions

STEP 7: Design the pile cap. (Some examples are given below.)

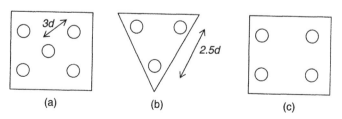

(a)   (b)   (c)

Figure 29.6   Pile caps. (a) Pile group with 5 piles. (b) Pile group with 3 piles. (c) Pile group with 4 piles.

Example

STEP 1: Assume 12-inch steel pipe piles.

STEP 2: Assume 3d center-to-center spacing between piles.

STEP 3: Assume the ultimate bearing capacity of a pile to be 100 tons.

STEP 4: Use a factor of safety of 2.5 to obtain the design pile capacity of 40 tons.

STEP 5: Assume a column load of 200 tons. Hence, five piles are required to carry the column load.

STEP 6: Use a square pile cap configuration as shown in Figure 29.6a.

30

Design Drawings and As-Built Drawings

30.1 Design Drawing Preparation

The design engineer should provide adequate information to the contractor and pile-driving inspector.

The following information is vital and should be provided without any ambiguity.

- Location of piles
- Allowable tolerance for the location (In some cases, piles may not be driven at the exact location indicated by the design engineer. In such cases, the contractor and pile-driving inspector should be given proper direction regarding the tolerance of change in location.) (*Example*: In the case of driving difficulties, the contractor may move the location of the pile no more than 6 in.)
- Location of borings and boring logs
- Contour map of the bearing stratum
- Easy to follow driving criteria

Information to the pile-driving inspector should be provided in an easily readable manner. A table as shown here should be provided to the pile-driving inspector so that he can take it to the field during pile driving.

Pile Group #	Nearby Borings	Depth to Bearing Stratum (ft)	Estimated Depth to Bearing Stratum at Pile Group	Engineer's Comments
1A	B5 B6 B10	B1 = 15 B6 = 25 B10 = 18	20 ft	Expect boulders at 15 ft
1B	B3 B9 B11	B3 = 35 B9 = 30 B11 = 25	32 ft	Nearest boring is 50 ft away. Soil conditions at this location could vary significantly.
1C	B1 B8 B9	B1 = 32 B8 = 30 B9 = 22	35 ft	

The following information is provided for the foundation plan (F-1) shown on the next page.

Horizontal centerlines are given letters A through D. (See the boxes on the left.)
Vertical centerlines are given numbers 1.0, 2.0, 2.1, 2.2, and 3.0.
Centerlines 1.0, 2.0, and 3.0 have columns from one end to the other.
Centerlines 2.1 and 2.2 have only one or two columns.
The letter G means grade beam.
The letter T means tie beam.
X–X is a cross section across a grade beam.
Y–Y is a cross section across a tie beam.
Z–Z is a cross section across a column line.

The design engineer should provide all the necessary cross sections for the contractor to perform the work.

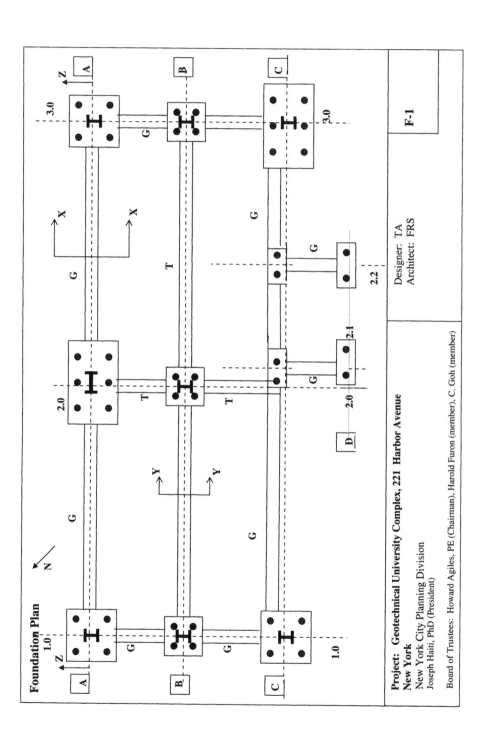

Foundation Plan

Project: **Geotechnical University Complex, 221 Harbor Avenue**
New York
New York City Planning Division
Joseph Haiti, PhD (President)

Designer: TA
Architect: FRS

F-1

Notes:

1. All material, construction methods, inspections, tolerances, fabrication and installations should conform to the local building code.

2. All backfill material shall be approved by the Engineer prior to use.

3. Compaction of backfill material should be done to with layers of 12 inches or less. Modified Proctor compaction of 95% should be achieved. All compaction tests shall be conducted by certified technicians belonging to the inspection company chosen for this project.

4. The contractor shall follow all safety requirements promulgated by OSHA and other local jurisdictions.

5. The contractor shall obtain utility maps from utility companies and markout utilities prior to start of any excavation. If it is possible the contractor shall have utility company representatives markout utility locations in the ground.

6. The contractor shall obtain prior approval from the Engineer prior to any excavation within 20 feet of an existing building.

7. The contractor shall verify all dimensions and in doubt, should consult the Engineer through a formal RFI. (Request for information).

8. The contractor shall prepare an as built plan indicating exact location of the pile with relative to the design location. Pile locations shall be surveyed by a licensed surveyor.

9. The contractor shall seek permission from the Engineer prior to filling the pile with concrete.

Project: **Geotechnical University Complex, 221 Harbor Avenue** **New York**	Architects & Engineers: **Ruwan Rajapakse Incorporated** 111 Chorkes Street, Yonkers, NJ	Drawing Title:Foundation n Submission: 100% Design No: D-1231	**F-2**
New York City Planning Division Joseph Haiii, PhD (President)	Project Manager: A. Saterta, PE Chief Engineer: B. Galol, PE Chief Architect: B. Nasaroy,	Rev. Date 2/11/02	Sh. 2 of 9
Board of Trustees: Howard Agiles, PE (Chairman), Harold Furon (member), C. Goh (member)			

Elevations:

1. Top of pile cap El. = 10 ft. (As per ABC Grid)
2. Top of grade beam El. = 14 ft.
3. Top of pier El. = 14 ft.
4. Top of first floor slab elevation = 15.00

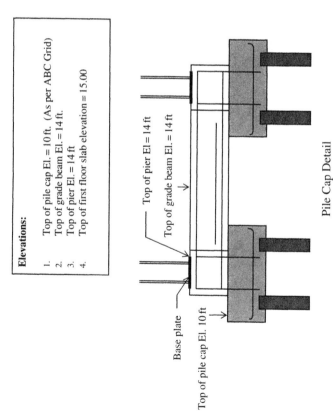

Top of pier El = 14 ft

Top of grade beam El = 14 ft

Base plate

Top of pile cap El. 10 ft

Pile Cap Detail

Project: Geotechnical University Complex, 221 Harbor Avenue New York	**Architects & Engineers:** Ruwan Rajapakse Incorporated 111 Charles Street, Yonkers, NJ	Drawing Title: Foundation Plan Submission: 100% Design No: D-1231
New York City Planning Division Joseph Haiti, PhD (President) Board of Trustees: Howard Agiles, PE (Chairman), Harold Furon (member), C. Goh	Project Manager: A. Saterta, PE Chief Engineer: B. Galol, PE Chief Architect: B. Nasaroy.	Rev. Date: 2/11/02 **F-3** **Sh. 3 of 9**

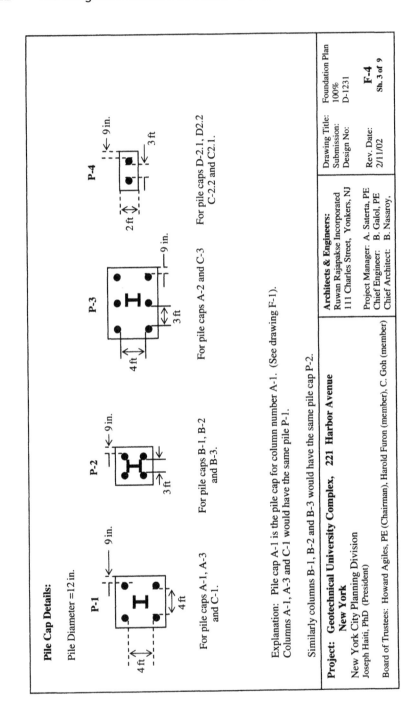

Pile Cap Details:

Pile Diameter =12 in.

P-1

⊢ 9 in.

4 ft

4 ft

For pile caps A-1, A-3 and C-1.

P-2

⊢ 9 in.

3 ft

For pile caps B-1, B-2 and B-3.

P-3

⊢ 9 in.

4 ft

3 ft

For pile caps A-2 and C-3

P-4

⊢ 9 in.

2 ft

3 ft

For pile caps D-2.1, D2.2 C-2.2 and C2.1.

Explanation: Pile cap A-1 is the pile cap for column number A-1. (See drawing F-1).
Columns A-1, A-3 and C-1 would have the same pile P-1.

Similarly columns B-1, B-2 and B-3 would have the same pile cap P-2.

Project: Geotechnical University Complex, 221 Harbor Avenue **New York** New York City Planning Division Joseph Haiti, PhD (President)	**Architects & Engineers:** Ruwan Rajapakse Incorporated 111 Charles Street, Yonkers, NJ Project Manager: A. Saterta, PE Chief Engineer: B. Galol, PE Chief Architect: B. Nasaroy,	Drawing Title: Foundation Plan Submission: 100% Design No: D-1231 Rev. Date: **F-4** 2/11/02 **Sh. 3 of 9**

30.2   As-Built Plans

After construction of piles, it is necessary to prepare an as-built pile location plan. Typically, a licensed surveyor develops the as-built pile location plan.

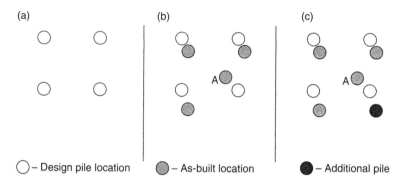

Figure 30.1   Pile locations

- Figure 30.1a shows the design locations of a pile group. Figure 30.1a shows actual locations where the piles were driven. As one can see, the pile group is not symmetrical. Pile A would be overloaded beyond the design capacity. Hence, the engineer may decide to add another pile to relieve the load on pile A.

30.2.1   Batter Information

- In reality, a large number of piles would have a batter, after they have been driven. Piles that have a batter larger than the allowable limit should be identified. The batter angle of piles should be provided to the engineer.

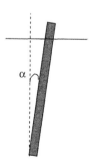

Figure 30.2   Batter angle of piles

Typically, batter angle (the angle measured from the vertical) is measured and provided to the design engineer. If the batter angle is excessive, allowable capacity of the pile needs to be reduced.

If the allowable pile capacity is q and the pile has a batter angle of α, new allowable pile capacity is given by $= q \times \cos(\alpha)$.

- Pile batter will reduce the axial capacity of piles. If piles have excessive batter, then new piles need to be added.

APPENDIX

Appendix A
Soil Mechanics Relationships

A.1 SPT (N) Value and Friction Angle (φ)

Young's Modulus of Clay Soils
Shear Modulus
SPT-CPT Correlations

SPT (N) Value and Friction Angle

Soil Type	SPT (N_{70} value)	Consistency	Friction Angle (φ)	Relative Density (D_r)
Fine sand	1–2	Very loose	26–28	0–0.15
	3–6	Loose	28–30	0.15–0.35
	7–15	Medium	30–33	0.35–0.65
	16–30	Dense	33–38	0.65–0.85
	<30	Very dense	<38	<0.85
Medium sand	2–3	Very loose	27–30	0–0.15
	4–7	Loose	30–32	0.15–0.35
	8–20	Medium	32–36	0.35–0.65
	21–40	Dense	36–42	0.65–0.85
	<40	Very dense	<42	<0.85
Coarse sand	3–6	Very loose	28–30	0–0.15
	5–9	Loose	30–33	0.15–0.35
	10–25	Medium	33–40	0.35–0.65
	26–45	Dense	40–50	0.65–0.85
	<45	Very dense	<50	<0.85

Source: Bowles (2004).

A.2   Young's Modulus of Clay Soils

- Laboratory tests for Young's modulus of clay soil are typically done in the triaxial test apparatus. These tests can be done under either drained or undrained conditions.

- In most cases, the difference between two modulus values is not great.

Young's Modulus—E_s(MN/sq m)

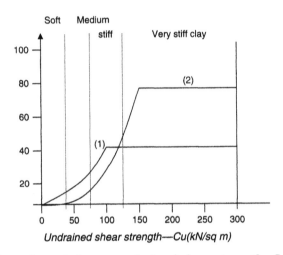

Figure A.1   Young's modulus vs. undrained shear strength. Curve 1: Driven Piles (*Poulos, 1989*). Curve 2: Bored Piles (*Poulos, 1989*).

Design Example: Find the Young's modulus value of a bored pile constructed in clay, which has an undrained shear strength of 100 kN/sq m.
Solution: Use curve 2 for bored piles.
$E_s = 25$ MN/sq m

Reference

Poulos, H.G., "Pile Behavior—Theory and Application," *Geotechnique*, 366–403, 1989.

Modulus of Elasticity of Sands and Clays

Soil Type	Modulus of Elasticity (psi)
Very soft clay	50–400
Soft clay	250–600
Medium clay	600–1,200
Hard clay	1,000–2,500
Sandy clay	4,000–6,000
Silty sand	1,000–3,000
Loose sand	1,500–3,500
Dense sand	7,000–12,000
Dense sand and gravel	14,000–28,000

Source: Arpad Kezdy (1975).

Modulus of Elasticity of Pile Materials

Pile Type	Modulus of Elasticity (psi)
Wood	1,200,000–1,500,000
Concrete	2,800,000–4,000,000
Steel	30,000,000

Source: Arpad Kezdy (1975).

Reference

Winterkorn, H.F., and Fang, H.Y., *Foundation Engineering Handbook*, article by Arpad Kezdy, "Deep Foundations," Van Nostrand Reinhold, New York, 1975.

A.3   Shear Modulus

A.3.1   Shear Modulus of Sandy Soils

The following approximate equation is provided by Lee et al. (1988) to find the shear modulus in sandy soils.

$G_s = 200\ \sigma_v'$ (for sands); $\sigma_v' =$ vertical effective stress

$G_s = 150\ C_u$ (for clays); G_s would have the same units as C_u.

Reference

Lee, S.L., et al. "Rational Wave Equation Model for Pile Driving Analysis," *J. of ASCE Geotechnical Eng.*, March 1988.

A.3.2 SPT-CPT Correlations

In the United States, the Standard Penetration Test (SPT) is used extensively. The CPT (Cone Penetration Test) is popular in Europe.

Europeans have developed many pile design methods using CPT data.

The following correlation between SPT and CPT can be used to convert CPT values to SPT number.

SPT-CPT Correlations for Clays and Sands

Soil Type	Mean Grain Size (D_{50}) (measured in mm)	Q_c/N
Clay	0.001	1
Silty clay	0.005	1.7
Clayey silt	0.01	2.1
Sandy silt	0.05	3.0
Silty sand	0.10	4.0
Sand	0.5	5.7
	1.0	7.0

Source: Robertson et al. (1983).

Q_c = CPT value measured in bars (1 bar = 100 Kpa)
N = SPT value

D_{50} = Size of the sieve that would pass 50% of the soil

Example: SPT tests were done on a sandy silt with a D_{50} value of 0.05 mm. The average SPT (N) value for this soil is 12. Find the CPT value.

Solution: From the preceding table, for sandy silt with a D_{50} value of 0.05 mm

$$Q_c/N = 3.0$$

$N = 12$; hence, $Q_c = 3 \times 12 = 36$ bars $= 3,600$ Kpa.

Standard CPT Device
A standard cone has a base area of 10 sq cm and an apex angle of 60°.

Standard SPT Device
Donut hammer with a weight of 140 lbs and a drop of 30 in. Donut hammers have an efficiency of 50 to 60%.

Reference

Robertson, P.K., et al., "SPT-CPT Correlations," *J. of ASCE Geotechnical Engineering*, November 1983.

Specific Gravity (G_s)

Soil	Specific Gravity
Gravel	2.65–2.68
Sand	2.65–2.68
Silt (inorganic)	2.62–2.68
Organic clay	2.58–2.65
Inorganic clay	2.68–2.75

Poisson's Ratio

Soil	Poisson's Ratio (μ)
Saturated clay	0.4 to 0.5
Unsaturated clay	0.1 to 0.3
Sandy clay	0.2 to 0.3
Silt	0.3 to 0.35
Dense sand	0.2 to 0.4
Coarse sand	0.15
Fine sand	0.25
Rock	0.1 to 0.4
Concrete	0.15

Size Ranges for Soils and Gravels

Soil	Size	Size	Comments
Boulders	6 in. or larger	150 mm or larger	
Cobbles	3 to 6 in.	75 mm to 150 mm	
Gravel	0.187 in. to 3 in.	4.76 mm to 75 mm	Greater than #4 sieve size
Sand	0.003 in. to 0.187 in.	0.074 mm to 4.76 mm	Sieve #200 to sieve #4
Silt	0.00024 in. to 0.003 in.	0.006 mm to 0.074 mm	Smaller than sieve #200
Clay	0.00004 in. to 0.00008 in.	0.001 mm to 0.002 mm	Smaller than sieve #200
Colloids	less than 0.00004 in.	Less than 0.001 mm	

Sieve Sizes

U.S. Sieve No.	Size (mm)	British Sieve No.	Size (mm)
4	4.76		
10	2.00	8	2.057
20	0.841	16	1.003
30	0.595	30	0.500
40	0.420	36	0.422
50	0.297	52	0.295
60	0.250	60	0.251
80	0.177	85	0.178
100	0.149	100	0.152
200	0.074	200	0.076
270	0.053	300	0.053

Note
Sizes greater than #4 sieve are considered to be gravel.
Sands—#4 to #200
Silts and Clays—smaller than #200

Index

Printed and bound by CPI Group (UK) Ltd, Croydon, CR0 4YY

03/10/2024

01040435-0007